SpringerBriefs in Electrical and Computer Engineering

More information about this series at http://www.springer.com/series/10059

Joseph Suresh Paul · Subha Gouri Raveendran

Understanding Phase Contrast MR Angiography

A Practical Approach with MATLAB Examples

 Springer

Joseph Suresh Paul
Medical Image Computing
    and Signal Processing Group
Indian Institute of Information Technology
    and Management-Kerala
Trivandrum
India

Subha Gouri Raveendran
Medical Image Computing
    and Signal Processing Group
Indian Institute of Information Technology
    and Management-Kerala
Trivandrum
India

Additional material to this book can be downloaded from http://extras.springer.com.

ISSN 2191-8112   ISSN 2191-8120 (electronic)
SpringerBriefs in Electrical and Computer Engineering
ISBN 978-3-319-25481-4   ISBN 978-3-319-25483-8 (eBook)
DOI 10.1007/978-3-319-25483-8

Library of Congress Control Number: 2015957051

This Springer imprint is published by SpringerNature
The registered company is Springer International Publishing AG Switzerland

Preface

This book discusses the basic concepts of MRI leading to PC-MRA. An intuitive understanding of PC-MRA concepts is provided through simulation techniques using the extended form of Bloch equation. For completeness, quantitative flow measuring techniques and post-processing using statistical models are also included as separate chapters. Implementation details of important techniques discussed in this book are provided in the form of MATLAB codes.

PC-MRA is one of the non-contrast MRA techniques using the idea that blood flow velocities can be encoded by phase. This was first developed by Paul R. Moran in the early 1980s. Moran analyzed the phase effects on stationary and moving spins subjected to a pair of bipolar gradients. A stationary spin subjected to such a gradient pair will experience no net phase shift, but a moving spin will have a net phase shift proportional to its velocity. Two spins flowing at the same speed but in opposite directions will have equal but opposite phase shifts, and by measuring changes in phase, the velocity can be computed.

PC-MRA is based on use of bipolar gradients that create phase shifts of moving spins proportional to their velocities. The key applications include flow measurements, Cine CSF flow studies, and venography. The degree of sensitivity to slow or fast flows is determined by the amplitude, duration, and spacing of bipolar gradients, which is controlled by parameter VENC—velocity encoding. Simulation experiments outlined in this book provide a sound understanding of the encoding strategy and enable the reader to apply this knowledge in acquisition and post-processing methods.

Acknowledgments

Some sections of this book are based on previous articles: "Computer Simulation of Magnetic Resonance Angiography Imaging: Model Description and Validation," Plos one 9. (2014) and "*In Silico* Modeling of Magnetic Resonance Flow Imaging in Complex Vascular Networks," IEEE transactions on medical imaging, 33: 11 (2014).

Contents

Chapter 1
Introduction to MR Imaging

Abstract This chapter provides a quick introduction to the basics of MRI, adequate for understanding of the terminology and concepts used in Phase-Contrast MR Angiography (PC-MRA). Chapter begins with a description of signal generation, followed by an explanation for the need of spatial encoding necessary for derivation of the Fourier imaging approximation. A brief introduction to the magnetization dynamics using Bloch equation, with extension to small tip angle approximation is also provided. The chapter concludes with an introduction to the basic principle of flow measurement using PC-MRA, and the need for additional flow encoding gradients.

Keywords Spin physics · RF excitation · Bloch equation · Spatial encoding · PC-MRA

1.1 Magnetic Resonance Imaging

Magnetic Resonance Imaging (MRI) is a non-invasive method of mapping the tissue in terms of its spatial density of protons. Since the proton density varies with the type of tissue, imaging contrast is achieved to differentiate tissue types based on variations in their proton density. The imaging contrast can be easily maneuvered by utilizing the dependence of received signal intensity on the physical parameters of proton density, longitudinal relaxation time (T_1), and transverse relaxation time (T_2). For example, chemical and structural changes of tumors over time directly affect the signal intensity on MR images, providing information about their age. Unlike many other medical imaging modalities, the contrast in an MR image is strongly dependent upon the way the image is acquired. By adding Radio Frequency (RF), or gradient pulses, and by careful choice of timings, it is possible to highlight different components in the object being imaged.

The MR Image is constructed by placing the patient inside a large magnet, which induces a relatively strong external magnetic field (B_0). This causes the nuclei of

© The Author(s) 2016
J. Suresh Paul and S. Gouri Raveendran, *Understanding Phase Contrast MR Angiography*, SpringerBriefs in Electrical and Computer Engineering, DOI 10.1007/978-3-319-25483-8_1

many atoms in the body, including Hydrogen, to align them with the magnetic field. With application of RF signal, energy is released from the area being imaged, in the form of a time-varying voltage. The detector demodulates the RF components, and stores the discrete samples in a Fourier space, known as k-space. The MR image is then obtained by performing an inverse Fourier transform on the acquired k-space.

Using slice-selection gradients, the imaging plane can be optimized for the anatomic area being studied. Flow-sensitive pulse sequences and MR Angiography (MRA) yield data about blood flow, as well as displaying the vascular anatomy. In conventional MR imaging, the spins are considered to be stationary throughout the imaging process. However, this does not hold true in the case of flow and vascular imaging. This is particularly applicable to the situation of imaging a slice containing blood vessels. A critical problem with flow imaging is that the excited spins in a vessel can flow out of the slice by the time readout is performed. Since the unexcited spins have flown in during readout, the image formed will not have contributions from MR signals originated from the vessel. Consequently, measurements of flow would require some form of spatial encoding that is flow sensitive. This is done by applying a magnetic field gradient along the direction in which flow is to be measured. A large enough gradient amplitude can dephase the stationary as well as moving spins depending on their position along the gradient. The gradient when reversed, will completely rephase only the stationary spins. Spins that have moved will not be completely rephased. For uniform flow, the phase difference and flow are directly related to the time delay between the forward and reverse gradients.

Spins that are moving in the same direction as a magnetic field gradient develop a phase shift that is proportional to the spin velocity. This is the basis of Phase-Contrast MRA (PC-MRA). In PC-MRA pulse sequence, the spin velocities are encoded using bipolar gradients (two gradients with equal magnitude but opposite direction). Following the gradient application, stationary spins do not undergo a net change in phase. Due to varying spatial position, the moving spins will experience a different magnitude of the second gradient compared to the first. As a result, the moving spins attain a net phase shift during readout. The resultant phase information can be used directly to determine the spin velocity.

Understanding the key physical concepts in MRI is a prerequisite for analyzing the theory and post processing methods used in PC-MRA. The successive chapters in this book are organized to orient the reader towards systematically building up the knowledge base needed to understand the technical aspects through a series of methods which outline the interface between flow and image formation process. This chapter provides the theoretical prelude to understand the key MRI concepts from a technical perspective. Chapter 2 highlights the fundamental approach to generate synthetic MR images from data consisting of relevant physical parameters, and geometry using known MR sequences. Chapter 3 describes the theory of PC-MRA together with details of methods used to derive flow information from raw data. Chapter 4 enables the reader to extend the image generation ideas presented in Chap. 2 for application to PC-MRA image generation. Chapter 5 summarizes the statistical methods for post-processing the flow information derived from PC-MRA. Links are provided in each chapter, for access to relevant MATLAB codes for simulation, pre-processing, PC-MRA image generation, and statistical analysis.

1.2 MRI Physics

1.2.1 Spin Physics

MRI uses large magnetic fields to magnetize the proton spins of human tissue. By changing the RF pulses and gradients, different contrasts can be obtained which enhance different aspects of the MR image. The way in which the RF pulses and gradients are altered to create an image is called a sequence.

All protons are spinning, and therefore, creates a tiny magnetic field, referred to as a "magnetic moment". The Pauli's exclusion principle [1] demands that no two subatomic particles in the same atom exist simultaneously in the same state. If the nucleus has an even number of protons, each proton will be accompanied by another of exactly the opposite spin, and the two magnetic moments will cancel each other.

Nuclear particles do not act as classical particles. A classical particle spinning in a magnetic field will, according to Maxwell's equation, radiate electromagnetic energy. These nuclear particles, even though they possess a magnetic moment aligned with the external field, do not emit, until stimulated by an RF pulse described next. Thus, only if a nucleus has an odd number of protons, will it possess a net magnetic moment. Placing this nucleus in a strong, constant magnetic field will cause it to tend to align with the field.

Understanding that we are only using an analogy, we will continue to refer to the quantum-mechanical property [2, 3] as "spin," and say that the nucleus now spins in such a way that the magnetic moment aligns with the external field. For simplification, the spins are often looked upon as spin packets which represent all the spins in a certain volume.

When there is magnetic dipole (or a current loop, constituted by orbiting electron) in applied magnetic field, there are various things which can happen: the magnetic dipole may align in same direction, opposite direction or make some angle with the field. When it makes some angle, it may draw some energy from the field and precess about the applied field with some precession frequency. Now spin angular momentum of a particle also acts as a dipole which can precess independently. Usually Larmor precession term is reserved for dipole moment. Now we must consider the rate of spin. Once the external field has been applied, the spins precess with Larmor frequency

$$\omega_0 = \gamma B_0 \tag{1.1}$$

Under the presence of magnetic field along z direction, the spin packets precess at a single frequency while they are not in phase. The resultant net magnetic field is pointed in z direction. Net magnetization M is the vector sum of nuclear magnetic moments μ_i produced by the spin packets. This is expressed as

$$M = \sum_{n=1}^{Ns} \mu_n \tag{1.2}$$

1.2.2   RF Excitation

The signals generated due to static magnetic field contain frequencies in the RF range, and are not useful for derivation of tissue-related information. Hence the received signals are first amplitude demodulated to extract the desired information. Considering the mathematical description of signals, this is equivalent to referring the spin dynamics in a rotating coordinate system, rotating at the RF Larmor frequency [4, 5]. The signal representation in the rotating frame of reference will contain only the intermediate frequencies in the kHz range. Consider the coordinate system for observation of the magnetization to be rotating at a frequency ω_0. The coordinate system is rotated about the z-axis in the same direction that the local magnetization M rotates about B_1. The "laboratory" frame of reference is the usual frame of reference with coordinates (x, y, z). The "rotating" frame of reference has coordinates (x', y', z'). For a rotating frame, the coordinate axes are transformed using

$$
\begin{aligned}
i' &= i\cos(\omega_0 t) - j\sin(\omega_0 t)\\
j' &= i\cos(\omega_0 t) + j\sin(\omega_0 t)\\
k' &= k
\end{aligned}
\tag{1.3}
$$

Consider an RF field B_1 (perpendicular to static field) applied for a short duration, to rotate the spin away from the initial B_0 alignment. The RF field is produced by a coil system separate from the static field source, called transmit field. When $B = B_0$ k, the field (B_{eff}) in the rotating frame will be effectively zero. Such rotation leaves the classical moment precessing at an angle around the original static field. The x-y components of the magnetization in rotating frame is given by

$$M_{xy.rot}(t) = M_{xy}(t) \exp(i\omega_0 t) \tag{1.4}$$

With the magnetization vector referenced in the static frame represented by $M = [M_x, M_y, M_z]$ and that in the rotating frame by $M_{rot} = [M_{x,rot}, M_{y,rot}, M_{z,rot}]$,

$$M_{xy \cdot rot}(t) = M_{x,rot} + iM_{y,rot}$$
$$= M_{xy} \exp(i\omega_0 t) \tag{1.5}$$

The longitudinal magnetization remains same in the rotating frame. For an RF field B_1 with frequency ω_1 applied along the x-direction,

$$\mathbf{B}_1 = B_1 \exp(-j\omega_1 t) \tag{1.6}$$

Since $\omega_1 = \omega_0$ at resonance, the total applied **B**-field can be represented as

$$\mathbf{B} = \mathbf{B}(t) = \begin{bmatrix} B_1 \cos(\omega_0 t) \\ -B_1 \sin(\omega_0 t) \\ B_0 \end{bmatrix} \tag{1.7}$$

In the rotating frame, the rate of change of magnetization is

$$\frac{dM_{rot}}{dt} = M_{rot} \times \gamma \mathbf{B}_{1eff} \tag{1.8}$$

where $\mathbf{B}_{1eff} = B_1 i'$ is the effective magnetic field in the rotating frame. Substituting for $\mathbf{B}_{1eff}$, the matrix version of (1.8) will be

$$\frac{d\mathbf{M}_{rot}}{dt} = \mathbf{M}_{rot} \begin{bmatrix} 0 & 0 & 0 \\ 0 & 0 & \gamma B_1 \\ 0 & -\gamma B_1 & 0 \end{bmatrix} \tag{1.9}$$

Using the initial condition $\mathbf{M}_{rot}(0) = m_0 k$, solution of (1.8) yields

$$\mathbf{M}_{rot}(t) = \begin{bmatrix} 0 \\ m_0 \sin(\gamma B_1 t) \\ m_0 \cos(\gamma B_1 t) \end{bmatrix} \tag{1.10}$$

Since B_{1eff} is applied along the x' axis, $\mathbf{M}_{rot}$ will precess around x' in the z-y' plane. In the rotating frame, the trajectory of the magnetization vector at any given time point during excitation is shown in Fig. 1.1.

Net magnetization M can be flipped to an angle α by applying magnetic field B_1 for a time period T such that

$$\alpha = \gamma B_1 T. \tag{1.11}$$

If the magnetization is to be flipped into the transverse (x-y) plane, the B_1 field is applied for a period of time and then removed. For a flip angle of $\alpha = \pi/2$,

Fig. 1.1 The trajectory of
M in the presence of B_0 and
B_1 in the reference frame

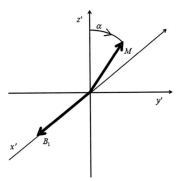

$$T = \frac{\pi}{2\gamma B_1}. \tag{1.12}$$

As noted from (1.8) the Bloch equation governs the precession of a nuclear magnetic moment $M(x) = ((M_x(x), M_y(y), M_z(z))^T$ about a time varying external magnetic field $\omega(t) = (\omega_x(t), (\omega_y(t), (\omega_z(t))^T$. In the presence of a gradient field, Bloch equation relates gradient fields and RF magnetic fields to the time derivative of magnetization. In a frame rotating at Larmor frequency, the Bloch equation is

$$\frac{dM(r,t)}{dt} = \gamma \begin{bmatrix} 0 & G(t) \cdot r & -B_{1,y}(t) \\ -G(t) \cdot r & 0 & B_{1,x}(t) \\ B_{1,y}(t) & -B_{1,x}(t) & 0 \end{bmatrix} M(r,t) \tag{1.13}$$

where γ is the gyromagnetic ratio, gradient fields $G(t) = ((G_x(t), G_y(t), G_z(t))^T$, and the RF magnetic field B_1 in (1.6) becomes a function of t with components $B_{1,x}(t)$ and $B_{1,y}(t)$ respectively. Here, the RF fields are assumed to be spatially uniform and (1.13) can be used to determine the magnetization state following RF excitation provided, an initial magnetization state, RF pulse and gradient waveforms are given.

1.2.3 Relaxation

The transverse component of the precessing magnetization M_{xy} can be detected by placing coils in the x-y plane. The rotating M_{xy} induces electromotive force in the coil. The magnetization vector (M) continuously precesses in the x-y plane under the absence of relaxation effects. But the relaxation mechanisms will cause dephasing of transverse magnetization, and a build-up of the longitudinal magnetization. The resulting process is called Free Induction Decay (FID). The relaxation process is divided into two independent processes which happens simultaneously. T_1 relaxation describes change in magnetisation vector in z-direction and T_2 relaxation describes changes in the x-y plane [5].

T_1 relaxation applies to protons in the volume following the 90° excitation pulse. During relaxation process, the spin will transfer the energy which are gained from RF pulse to the environment. Larger molecules exhibit slower relaxation and vice versa. T_1 values are determined from the relation between Larmor frequency and natural frequency of the molecules. The values are small when both frequencies are closer. The process of T_1 relaxation is mathematically represented using exponential growth, described using

$$M_z(t) = M_0 \left(1 - e^{-t/T_1} \right) \tag{1.14}$$

The second relaxation process, the transverse relaxation, has its origin in the interaction of the spins with each other and the magnetic properties of their environment. The magnetic moments of neighboring spins trigger magnetic field fluctuations. Through these field inhomogeneities, the Larmor frequencies of spins tend to differ by small amounts, so that the spins lose their phase coherence. The dephasing of spins results in decrease of the transversal magnetization. The duration of this process is described by the time constant T_2, following which the transverse magnetization $M_{xy}(t)$ has decreased to about 37 % of its initial value. The rate of dephasing is different for each tissue. Fat tissue will dephase quickly, while water will dephase much slower. The transverse relaxation is mathematically represented using

$$M_{xy} = M_{xy}(0)e^{-t/T_2} \tag{1.15}$$

In reality, the dephasing is even more pronounced due to external field inhomogeneities, leading to the effective transversal relaxation time T_2^* ($T_2^* < T_2$). Mathematically, this effective relaxation rate is given by

$$\frac{1}{T_2^*} = \frac{1}{T_2} + \frac{1}{T_2'} \tag{1.16}$$

where T_2 is related to non-reversible spin-spin relaxation and T_2' refers to the reversible relaxation due to static field inhomogeneities. The two constants T_1 and T_2 are tissue-specific, where usually $T_1 > T_2$. These relaxation times can be used in MR imaging to generate different tissue contrasts in the images.

1.3 Signal Generation

The received signal $s(t)$ is composed of contributions from all spins possessing transverse magnetization in the imaged volume. This is mathematically expressed using

$$s(t) = \int_{vol} M_{xy}(r,t)dV = \int_z \int_y \int_x M_{xy}(x,y,z,t)dxdydz \qquad (1.17)$$

In the presence of transverse relaxation, the received signal equation becomes

$$s(t) = \int_{vol} M_{xy}(x,y,z,t)e^{-t/T_2(x,y,z)}dV \qquad (1.18)$$

In any MR imaging sequence, the maximum signal intensity is obtained at $t = TE$ (echo time). This is achieved by signal refocusing using either spin or gradient echo formation (see Chap. 2 for further information). Either forms of refocusing is performed in the presence of external gradient fields, described in later sections. The signal received in the absence of refocusing is referred to as free induction signal. In quadrature reception, the FID signals are acquired in the *in-phase* (*I*) and *quadrature* (*Q*) coils upon application of a 90° RF pulse in the phase-encode direction as shown in Fig. 1.2.

Imaging application often requires selection of a particular slice in which the applied gradient fields cause changes in spin precession. Slice selection is accomplished using combination of an RF excitation pulse and a slice selection gradient pulse. For sagittal and coronal sections, the slices are selected in the x and y directions respectively. By convention, the laboratory frames in the scanner are set with the x-direction from Left ear-to-Right ear (LR), and the y-direction from Anterior-to-Posterior (AP). Axial scanning requires slice selection gradients to be

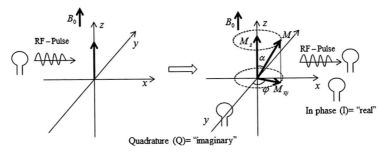

Fig. 1.2 After RF-pulse application, the magnetization M is tilted by the flip angle α from the equilibrium along B_0 direction and precesses with the Larmor frequency about the z-axis, parallel to B_0. M_z is the longitudinal and M_{xy} is the transverse component of magnetization

applied in the z-direction, which is same as direction of the main field B_0. Assuming that a selective slice excitation is used for axial scan, the dependence of M_{xy} on z can be eliminated. Since the net magnetization $M_{xy}(r, 0)$ at $t = 0$, is directly proportional to the spin density $\rho(r)$, the signal contribution from a voxel at position r can be represented as

$$s(x, y) = \rho(x, y)f(v)\left(1 - e^{-TR/T_1}\right)e^{-TE/T_2} \tag{1.19}$$

where $f(v)$ represents signal modulation arising from fluid flow, and TR and TE denote the repetition time and echo time respectively for any MR sequence. In (1.19), the time independence of the local signal (on the *LHS*), as opposed to (1.18), is obtained by choosing the maximum signal at the refocusing time. Since TR and TE are image sequencing parameters, the above relationship illustrates how their values can be manipulated to provide different types of imaging contrast. In the absence of external field gradients, the time-independent form of received signal is the sum total of $s(x, y)$ from all locations in the imaging Field Of View (FOV). In the time-dependent form, this is same as $s(t)$ in (1.18). The imaging problem consists of inverting the signal equation for computing the proton density as a function of location in the imaging plane. Even with discrete samples of signals at different time points, it is impossible to invert (1.18) for computation of $\rho(r)$. It is for the sake of enabling the inversion, that the locally generated MR signals are spatially encoded using externally applied magnetic field gradients. The mathematical nature of encoding signals, and their role in obtaining a Fourier approximation [6–8] for imaging, are discussed further in the succeeding sections.

1.3.1 Spatial Encoding of MR Signal

By varying the magnetic field spatially, the spins at each location will experience different magnetic field strengths and precess with different Larmor frequencies. This is achieved by adding magnetic gradients to the B_1 field. Addition of a positive gradient (Fig. 1.3) in the z-direction results in increasing precession frequency with z. When applying an RF-pulse, only spins in a slice with matching precession frequency will flip down. This is called slice selection, and it is possible to vary the thickness of a slice by changing the bandwidth of the RF-pulse [5, 8].

Consider a matrix of spins (spin packets) representing a slice. Following application of an RF pulse, the received signal will be

$$s(t) = \iint \rho(x, y)\exp(-t/T_2)\exp(-i\omega_0 t)dxdy \tag{1.20}$$

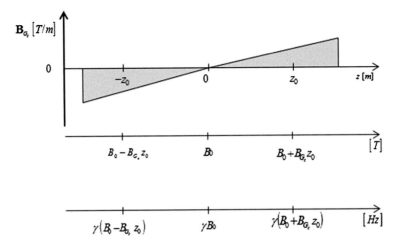

Fig. 1.3 The magnetic gradient applied in the z-direction

By adding a gradient field (G_y) in the y-direction for a short duration (τ), the spins will start to precess with same frequency, but different phase values depending on their y-position. Then the signal Eq. (1.20) takes the form

$$s(t) = \iint \rho(x,y) \exp(-t/T_2) \exp(-i(\omega_0 t + \varphi)) dx dy \qquad (1.21)$$

In (1.21), $\exp(-i\omega_0 t)$ can be removed by quadrature demodulation, and $\exp(-t/T_2)$ can be removed by assuming $T_2 \gg t$. The final step in the encoding sequence is to add a frequency-encoding gradient (G_x) in the x-direction. Considering the approximations described above, and inclusion of encoding gradients, the signal equation now becomes

$$s(t) = \iint \rho(x,y) \exp\big(-i\gamma(xG_x t + yG_y \tau)\big) dx dy \qquad (1.22)$$

With encoding, all the magnetization vectors have a different combination of frequency and phase, making it possible to localize the spatial origin of the spins.

1.3.2   2D Imaging

A pulse sequence for a 2D image acquisition must involve two gradients. Suppose a 90° pulse excites spins within a transverse slice at a chosen value of z, acquisition of spatially differentiated signal from the slice requires application of magnetic field gradients in the x and y directions (Fig. 1.4).

Fig. 1.4 The FID signal read
off in presence of phase and
frequency encoding

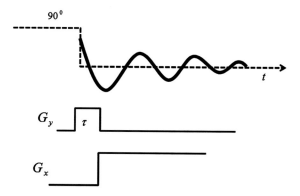

Before the start of acquisition, gradient G_y is applied for a fixed duration τ. Depending on the phase-encoding pulse amplitude G_y, the spins along y are dephased by

$$\varphi = \gamma y G_y \tau \tag{1.23}$$

Once the y gradient is turned off, the signal is sampled in the presence of frequency-encoding G_x. Since acquisition is performed in a discrete fashion, samples are acquired at discrete intervals of time Δt during each application of the frequency-encoding pulse. Number of samples acquired (N) during each application of G_x, determines the dimension of sampling plane in the frequency-encoding direction. Likewise, number of repetitions of G_x (M) determines the dimension of sampling plane in the phase-encoding direction. During each repetition, the amplitude of phase-encoding pulse G_y is changed at discrete interval. Imagining the sampling space to be filled with a set of uniformly spaced horizontal lines, acquisition of each horizontal line calls for an incremental change in the phase-encoding pulse amplitude. Starting from the central portion of the sampling plane, each line corresponds to a pulse amplitude of $G_y = m\Delta G_y$. Varying the value of phase-encoding index (m) from $(-M/2-1)$ to $(M/2)$, covers acquisition of lines from the bottom-half to top-half of the sampling plane. Equivalently, traversal of the entire sampling plane requires M step changes of phase-encode pulse amplitude. Thus, location in the sampling plane can be specified using the sample number during each repetition of the frequency-encoding pulse together with the phase-encoding index, pointing to the vertical location in the plane. Treating the central location as origin, the sampling index (n) takes values ranging from $(-N/2-1)$ to $(N/2)$. The variables k_x and k_y are defined in spatial frequency units (cm^{-1}) to specify the locations in sampling space. Assignment of spatial frequency unit to k_x and k_y are implicit in their relationship to the basic sampling variables Δt and ΔG_y, through the definitions:

$$k_x = \frac{\gamma G_x n \Delta t}{2\pi} \quad \text{and} \quad k_y = \frac{\gamma m \Delta G_y \tau}{2\pi} \tag{1.24}$$

Rewriting (1.22) using spatial frequency variables k_x and k_y,

$$s(k_x, k_y) = \iint \rho(x,y) \exp\big(-i(xk_x + yk_y)\big) dx dy \tag{1.25}$$

This is a familiar form of a 2D Fourier transform, and states that the acquired signal at each location in the sampling plane specified by the position vector $\mathbf{k} = [k_x, k_y]$, is obtained as the Fourier transform of the spin density function in the imaging plane. In MR literature, the sampling plane is referred to as the k-space. A k-space is different from a conventional Fourier space due to the fact that its samples are collected in a time-dependent fashion. For example, in MRI, the samples in each phase-encode line are acquired at discrete time intervals, and hence the collection of samples can be modelled as a discrete-time signal. Because of the Fourier Transform relation in (1.25), the spin density in the image plane can be computed as the inverse 2D Fourier transform of the signal samples in k-space. The arrangement of k-space samples in the $k_x - k_y$ plane is depicted in Fig. 1.5.

In a typical acquisition, the fourth quadrant is sampled after acquiring the first quadrant of $k_x - k_y$ plane, to complete the positive half-plane. This is achieved by allowing G_y to take on negative values. Similarly, the negative half-plane can be determined by allowing G_x to take on negative values.

However, sampling all quadrants requires lengthy acquisition. This is because each change in the gradient value requires a new signal to be acquired. If the image

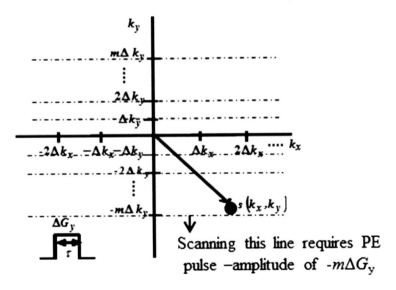

Fig. 1.5 Description of k-space

phase is not required, this problem can be overcome by sampling quadrants 1 and 3, or 2 and 4 only. After sampling any one of the two pairs of quadrants, the other quadrants can be obtained from Hermitian property of the Fourier transform [3]. It states that the Fourier transform of a real function (the image is a real function) has the Hermitian property:

$$s\left(-k_x, -k_y\right) = s^*\left(k_x, k_y\right) \tag{1.26}$$

1.3.3 Small Tip Angle Approximation

From (1.8), the effective magnetic field in the rotating frame B_{1eff} is computed as

$$\mathbf{B}_{1eff} = \mathbf{B}_{rot} + \frac{\omega_{rot}}{\gamma} \tag{1.27}$$

where $\omega_{rot} = \begin{bmatrix} 0 \\ 0 \\ -\omega \end{bmatrix}$. Letting $\mathbf{B}_{rot} = B_1(t)\mathbf{i} + B_0\mathbf{k}$, Then (1.27) becomes

$$\mathbf{B}_{1eff} = B_1(t)\mathbf{i} + \left(B_0 - \frac{\omega}{\gamma}\right)\mathbf{k} \tag{1.28}$$

If $\omega = \omega_0 = \gamma B_0$ at resonance, then $\mathbf{B}_{1eff} = B_1(t)\mathbf{i}$. Under this condition, the flip angle α can be expressed as

$$\alpha = \int_0^\tau \gamma B_1(t)dt \tag{1.29}$$

Upon application of a slice selection gradient G_z along z direction, the effective B_1 field under resonance condition becomes

$$\mathbf{B}_{eff} = B_1(t)\mathbf{i} + \gamma z G_z\mathbf{k} \tag{1.30}$$

Small Tip angle approximation [9] can be performed by considering the Fig. 1.6. For small α, $M_z = M_0 \cos \alpha \approx M_0$ and $M_{xy} = M_0 \sin \alpha \approx M_0\alpha$. At each location z in the rotating frame, the effective field is

$$B_1^e(t) = B_1(t) \exp(j\omega(z)t) \tag{1.31}$$

Using (1.3), the flip-angle at each z location is

Fig. 1.6 Small tip angle
approximation

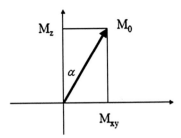

$$\alpha(t,z) = \gamma \int\limits_0^t \exp(j\omega(z)s)B_1(s)ds \qquad (1.32)$$

Using small angle approximation,

$$M_{xy}(t,z) \approx jM_0\alpha(t,z) = jM_0\gamma \exp(-j\omega(z)t) \int\limits_0^t \exp(j\omega(z)s)B_1(s)ds \qquad (1.33)$$

When the circularly polarized RF field is modulated by a rectangular pulse of amplitude B_1 and duration τ, application of small angle approximation yields

$$M_{xy}(\tau,z) = jM_0 \exp(-j\omega(z)\tau) \int\limits_0^\tau \exp(j\omega(z)s)\omega_1 rect\left(\frac{s+\tau/2}{\tau}\right)ds$$
$$\qquad (1.34)$$
$$= jM_0 \exp\left(-j\omega(z)\tau/2\right)\omega_1\tau \sin c\left(\frac{\gamma G_z\tau}{2\pi}z\right)$$

Small-tip-angle approximation to the Bloch equation simplifies small-tip-angle RF pulse design by a Fourier relationship between an RF pulse and magnetization pattern [10]. To derive this relationship, we make assumptions that (i) magnetization is at equilibrium at the beginning of the pulse (i.e.). $M(r,0) = (0, 0, M_0(r))^T$, where $M_0(r)$ is the equilibrium magnetization and (ii) the RF pulse $B_1(t)$ excites a small angle $\alpha < 30°$. These assumptions lead to the approximation that $M_z(t)$ remains constant during the RF pulse. With this approximation,

$$\frac{dM(r,t)}{dt} = \frac{dM_x(r,t)}{dt} + i\frac{dM_y(r,t)}{dt} \qquad (1.35)$$

From the Bloch equation with gradients in (1.13),

$$\frac{dM_x(r,t)}{dt} = G(t) \cdot rM_y(r,t) - B_{1,y}(t)M_0(r)$$

and   (1.36)

$$\frac{dM_y(r,t)}{dt} = -G(t) \cdot rM_x(r,t) + B_{1,x}(t)M_0(r)$$

Using (1.36) in (1.35),

$$\frac{dM(r,t)}{dt} = G(t) \cdot rM_y(r,t) - iG(t) \cdot rM_x(r,t) - iM_0(r)B_1(t)$$
$$= -iG(t) \cdot rM(r,t) - iM_0(r)B_1(t) \qquad (1.37)$$

Solution of this liner differential equation yields

$$M(r,T) = i\gamma M_0(r) \int_0^T B_1(t) e^{-i\gamma r \cdot \int_t^T G(t')t'} dt \qquad (1.38)$$

Introducing a spatial frequency variable $k(t)$ defined as a time reversed integration of gradient waveform $G(t)$, (1.38) can be expressed as

$$M(r,T) = i\gamma M_0(r) \int_0^T B_1(t) e^{ir \cdot k(t)} dt \qquad (1.39)$$

From (1.39), the excited pattern $M(r, T)$ is given by a Fourier integral weighted by the RF pulse at spatial frequency locations determined by the excitation k-space trajectory [11].

With small flip angle, contrast is less dependent on residual magnetization (M_{xy}). Since TR is often minimized, the flip angle can be used to adjust the contrast in rapid gradient echo sequences. In MRI pulse sequences, the same sequence of RF rotations and gradients is usually repeated many times, with magnetization changing the same way during each repetition. When the TR is long, the magnetization begins at equilibrium on each repetition. As TR is shortened to about $2 \times T_1$ or lower, incomplete T_1-relaxation may occur, and magnetization forms a "steady state" that depends on T_1 and TR. When TR is further shortened to about $2 \times T_2$ or lower, both T_1 and T_2 relaxation are incomplete and the steady-state signal depends on many factors. The basic gradient-echo sequence occurs when the net gradient area on each axis is zero over the sequence repetition. Therefore, the effect of gradients upon the steady state is neglected. The resulting dynamics of magnetization is called "steady-state free precession". The balanced steady state free precession signal profile is inverted so that at critical frequencies, the low flip angle signals are maximized. Small tip angle approximation is used for performing selective excitation which considers the pulse shape, pulse length and pulse frequency.

1.4 Phase Contrast Imaging

In phase contrast imaging, the contrast between flowing blood and stationary tissues is derived by manipulating the phase of the spin magnetization. From a physical perspective, the phase is indicative of the angle by which the local magnetization precesses from the time it is tipped into the transverse plane until the time it is detected. The process of enabling the phase to directly depend on the flow is called flow encoding. In order to encode the flow in each direction, a flow-encoding gradient is applied on each of the three gradient axes in separate TR intervals. In addition, a fourth non-flow-encoded acquisition is also performed. Data from the latter acquisition is subtracted from each of the three flow-encoded acquisitions to eliminate phase accumulation from sources other than velocity, such as field inhomogeneities. The need to acquire four acquisitions to encode flow in all directions lengthens the scan time. Nevertheless, subtraction results in high contrast between vessels and background.

The phase accrued by any spin in the presence of a gradient field can be represented mathematically in terms of a time integral of the product of its time-varying location along the gradient direction, and the instantaneous magnitude of the gradient waveform. Treating the motion of spins in a blood vessel to be linear, the integrand can be expanded as the sum of higher order motion terms weighted by the respective higher order gradient moments. A detailed mathematical description of this model is provided in Chap. 3.

In the presence of flow encoding gradients, the phase will be dependent on *zero*th and first order moments of the gradient waveforms. In the linear model, dependence of phase on initial position can be eliminated by adjusting the *zero*th order moment of the gradient waveform to zero. This is accomplished by use of bipolar gradients.

In practice, MRI phase measurements are contaminated by spurious (and often unknown) phase shifts that are constant in time but spatially varying [5–7]. Thus, each component of velocity in a phase contrast sequence is reconstructed from two separate acquisitions that are identical except for the polarity of the gradient waveform and its first moment M_1. The spurious phase component is removed when the two phase measurements are subtracted. After subtraction, the moving spins exhibit a phase difference that is proportional to the change in first moment. The required amount of first moment change is obtained by adjustment of the flow-sensitive gradient. For each setting, the velocity-encoding variable (VENC) is the value that produces a phase shift of 180°. Phase contrast sequences used in many vascular investigations are encoded in all three coordinate directions using a four-point encoding strategy. Figure 1.7 is schematic diagram of a four-point gradient echo phase contrast sequence for quantifying three components of velocity.

Slice selection is along the z axis, phase encoding is along the y axis, and signal read out is along the x axis. The number of phase encodes and sample read out determines the k-space matrix size. The first *TR* has a negative M_1 moment (the reference flow encodes) on all three axes. Each of the following *TR*s has a positive

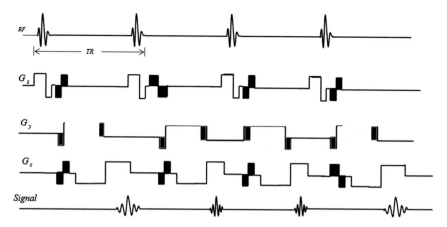

Fig. 1.7 A schematic diagram of a four-point gradient echo phase contrast sequence for quantifying three components of velocity. *Pure white* corresponds to positive velocity equal to the VENC, *pure black* corresponds to negative velocity equal to the VENC, and 50 % *gray* is zero velocity

M_1 moment on one axis and negative M_1 moments on the other two axes. Note that the phase encode level is the same for all four encodes and also that the flow encode lobes straddle the phase encode lobe to help minimize gradient switching. Four images are typically reconstructed for each three-component phase contrast image: magnitude, right-left velocity, anterior-posterior velocity, and superior inferior velocity.

The magnitude image has the appearance of a "typical" MR image, whereas the signal intensity in the velocity images is proportional to the velocity of motion in the given direction. Maximum intensity corresponds to positive velocity equal to the *VENC*, and minimum intensity corresponds to negative velocity equal to the *VENC*. Positive flow is generally in the direction of right to left, anterior to posterior, and superior to inferior. When measuring three components of blood flow velocity, it is advantageous to use separate velocity encoding values for each component to more accurately capture in-plane velocity patterns while avoiding aliasing of through-plane velocities [12].

References

1. Ben R (1989) An entirely fabulous account of the origin of the Pauli exclusion principle. J Chem Educ 66(12):983
2. Stephen GO (2012) Quantum mechanical review of MRI: clinical application. Lucid Res Lab 222(02):1–10
3. Montero G, Urena J, Dorado TC (2013) Dynamics of spin systems. Services editorial ADI Chapter 9, p 123

4. Matthew S (2012) The development and assessment of cardiac magnetic resonance imaging. Doctoral Thesis, University of St Andrews, chapter 1, vol 12, pp 10–15
5. Mark AB, Richard C (1999) Basic principles and applications of MRI, 3rd edn. Wiley, New Jersey
6. Odey S, Onwu Oluwaseun MD, Omotayo BA (2014) Physics and mathematics of magnetic resonance imaging for nano medicine. J Transl Med 3(1):17–30
7. Gallagher TA, Nemeth AJ, Hacein Bey L (2008) An introduction to the fourier transform: relationship to MRI. Am J Roentgenol 190:1396–1400
8. Gosta H, Hans K (1995) Signal processing for computer vision. Springer Science and Business media 4.2.9, pp 140–145
9. Pauly JM, Nishimura DG, Macovski A (1989) A k-space analysis of small-tip-angle excitation. Magn Reson Med 81:43–56
10. Grissom WA (2008) RF pulse design for parallel excitation in MRI. Doctoral Thesis, The University of Michigan
11. Pauly JM, Nishimura DG, Macovski A (1989) A linear class of large tip-angle selective excitation pulses. J Magn Reson 82:571–587
12. Charles AT, Mary TD (2004) Experimental and computational methods in cardiovascular fluid mechanics. Annu Rev Fluid University California Stanford 36:197–231

Chapter 2
Simulation Overview

Abstract MRI Simulator is based on numerical solution of the Bloch Equation. The numerical solution is obtained as a time update form of the magnetization vector. Parameters of this time update coefficients are shown to be related to the pulse sequence parameters and gradient amplitudes. The contents include application of the time update solution for simulation of Gradient Echo (GRE) sequences. Influence of T2* and susceptibility effects are also discussed.

Keywords Bloch equation · MRI simulator · T2* effect · SWI

Image quality of MRI, acquired using multiple phased array coils is often dictated by spatially varying noise and large geometry factors. For dynamic acquisitions such as cardiac and functional MRI (fMRI), the image quality is affected by subject movements, leading to geometric distortions. Further, in phase sensitive acquisition such as PC-MRA, the influence of noise mostly affects computations performed on phase images. The nature and influence of different types of noise sources can only be understood and remedial algorithms developed with the possibility of simulating the actual scanning process.

MR Simulator uses electrical and magnetic fields from electromagnetic field calculations to simulate realistic MR Images. The accuracy of simulator increases with the accuracy of model generated and its characteristics [1]. For example, it is possible to simulate spatial diffusion by convolving the time-dependent magnetization with a kernel representative of Gaussian diffusion. Apart from simulation of unrestricted diffusion, the convolution approach is also employed to model steady-state free precession [2] and gradient spoiling for steady-state sequences [3]. In an alternative form of simulation, the phenomenon of intra-voxel dephasing is simulated using Bloch-Torrey equation [4]. This involves an additional term in the Bloch equation that describes signal losses due to the diffusion process. The use of Bloch-Torrey equation for simulation of magnetization vectors allows more efficient simulation compared to that using spatial summation of isochromats. The technique is most applicable to simulating the effect of field perturbations,

© The Author(s) 2016
J. Suresh Paul and S. Gouri Raveendran, *Understanding Phase Contrast MR Angiography*, SpringerBriefs in Electrical and Computer Engineering, DOI 10.1007/978-3-319-25483-8_2

i.e. intra-voxel dephasing, but also for other typical imaging experiments and simulation of diffusion weighting [4]. Dephasing due to unrestricted motion such as blood flow could be simulated by incorporating voxel-dependent velocity and acceleration [5]. Bloch Equations, in general, compute temporal dynamics of the net magnetization. The Bloch MR Simulator can be implemented using realistic simulation of spatial field variation due to gradients and RF field effects in MRI.

2.1 Bloch Equation

Bloch equation provides a phenomenological description of the time dependence of magnetization. The solution of Bloch equation involves computation of magnetization vector $\vec{M} = (M_x, M_y, M_z)$ as a function of time. Magnetization along each cartesian direction is represented by the respective vector component along x, y, and z directions. For deriving the Bloch equation, the magnetic spin is assumed to be placed in a static magnetic field B_0, in the z direction. The magnetic moments orient either along field direction, or it's opposite. It causes the spins to precess at the Larmor frequency $\omega_0 = \gamma B_0$. Equation governing the dynamics of magnetization vector $\vec{M}$ in the static magnetic field is described using

$$\frac{d\vec{M}}{dt} = \gamma \vec{M} \times \bar{B}_0 \hat{k} \qquad (2.1)$$

where × represents cross product indicating that the rate of change of $\vec{M}$ is perpendicular to both $\vec{M}$ and $\bar{B}_0$. This implies that the initial spin precession is about the direction of the main magnetic field. Measuring the intensity of magnetization vector is practically impossible due to dominance of the external field B_0. In order to change the plane of precession, the magnetization $\vec{M}$ is tipped towards the transverse x-y plane. The factors affecting relaxation of the magnetization vector $\vec{M}$, are the relaxation time constants T_1 and T_2 [6, 7]. The spin-lattice relaxation time T_1 corresponds to the time required for the system to return to its equilibrium value after it has been exposed to a 90° tipping pulse. The spin-spin relaxation time T_2 depicts the time required for the tipped magnetization in the x-y plane to decay down to zero. The differential equation for magnetization in the presence of a magnetic field with relaxation terms can be combined to form a vector differential equation. In vectorized form, the Bloch equation becomes

$$\frac{d\vec{M}}{dt} = \gamma \vec{M} \times \vec{B}_{ext} + \frac{1}{T_1}(M_0 - M_z)\hat{k} - \frac{1}{T_2}\left(M_x\hat{i} + M_y\hat{j}\right) \qquad (2.2)$$

where $\vec{B}_{ext} = B_0\hat{k}$ and $\vec{M} = M_x\hat{i} + M_y\hat{j} + M_z\hat{k}$. Since the external field components are zero along x and y directions, the cross product term will be $M_yB_0\hat{i} - M_xB_0\hat{j}$. Substituting the cross product terms into (2.2) gives

$$\frac{d}{dt}\left(M_x\hat{i} + M_y\hat{j} + M_z\hat{k}\right) = \gamma\left(M_yB_0\hat{i} - M_xB_0\hat{j}\right) + \frac{1}{T_1}(M_0 - M_z)\hat{k}$$
$$- \frac{1}{T_2}(M_x\hat{i} + M_y\hat{j}) \tag{2.3}$$

Equating the coefficients of $\hat{i}, \hat{j}$ and $\hat{k}$ on both sides,

$$\frac{dM_z}{dt} = \frac{M_0 - M_z}{T_1} \tag{2.4}$$

$$\frac{dM_x}{dt} = \omega_0 M_y - \frac{M_x}{T_2} \tag{2.5}$$

$$\frac{dM_y}{dt} = \omega_0 M_x - \frac{M_y}{T_2} \tag{2.6}$$

2.1.1 Solution of Bloch Equation

Equation (2.4) is similar to the standard first-order differential equation, $\frac{dy}{dx} + P(x)y = Q(x)$; whose solution is directly obtained as $y\left(e^{\int p\,dx}\right) = \int Q e^{\int p\,dx} dx$. Substituting $P = \frac{1}{T_1}$ and $Q = \frac{M_0}{T_1}$, the solution of (2.4) takes the form

$$M_z(t) = M_0 + Ce^{-t/T_1} \tag{2.7}$$

Solving for C in terms of the initial longitudinal magnetization at $t = 0$, the time dependent form of the longitudinal magnetization will be

$$M_z(t) = M_z(0)e^{-t/T_1} + M_0\left(1 - e^{-t/T_1}\right) \tag{2.8}$$

To solve (2.5) and (2.6), the transverse magnetizations are assumed to be exponential in nature; each relaxing with time constant T_2.

$$M_x = m_x e^{\frac{-t}{T_2}}$$
$$and$$
$$M_y = m_y e^{\frac{-t}{T_2}} \tag{2.9}$$

From (2.9), we get $\frac{dM_x}{dt} = \frac{m_x}{T_2}$ and $\frac{dM_y}{dt} = \frac{m_y}{T_2}$. Substituting for M_x, M_y, $\frac{dM_x}{dt}$ and $\frac{dM_y}{dt}$ into (2.5) and (2.6),

$$\frac{dM_x}{dt} = \omega_0 m_y e^{\frac{-t}{T_2}} - \frac{m_x}{T_2} e^{\frac{-t}{T_2}} \tag{2.10}$$

and

$$\frac{dM_y}{dt} = \omega_0 m_x e^{\frac{-t}{T_2}} - \frac{m_y}{T_2} e^{\frac{-t}{T_2}} \tag{2.11}$$

Comparing (2.10) and (2.11) with (2.5) and (2.6),

$$\frac{dm_x}{dt} = \omega_0 m_y$$

and \qquad\qquad\qquad (2.12)

$$\frac{dm_y}{dt} = -\omega_0 m_x$$

Differentiating (2.12) once again,

$$\frac{d^2 m_x}{dt^2} = \omega_0 \frac{dm_y}{dt} \tag{2.13}$$

Substituting for $\frac{dm_y}{dt}$ from (2.12), (2.13) becomes

$$\frac{d^2 m_x}{dt^2} = \omega_0(-\omega_0 m_x)$$
$$= -\omega_0^2 m_x \tag{2.14}$$

Since (2.14) is a standard form of second-order differential equation, its solution is easily obtained as

$$m_x = C_1 \cos(\omega_0 t) + C_2 \sin(\omega_0 t) \tag{2.15}$$

Noting (2.9) through (2.15), the solution for time varying magnetization is

$$M_x = e^{\frac{-t}{T_2}}(C_1 \cos(\omega_0 t) + C_2 \sin(\omega_0 t))$$
and \qquad\qquad\qquad (2.16)
$$M_y = e^{\frac{-t}{T_2}}(-C_1 \sin(\omega_0 t) + C_2 \cos(\omega_0 t))$$

Considering the initial values of M_x and M_y to be $M_x(0)$ and $M_y(0)$, the unknown constants C_1 and C_2 can be evaluated. For a constant field $B_{ext} = B_0 \hat{k}$ in the longitudinal direction, the final form of solution of the Bloch equation will be

$$\vec{M}(t) = \begin{bmatrix} M_z(0)\mathrm{e}^{-t/T_1} + M_0\left(1 - \mathrm{e}^{-t/T_1}\right) \\ \mathrm{e}^{\frac{-t}{T_2}}(M_x(0)\,\cos(\omega_0 t) + M_y(0)\,\sin(\omega_0 t) \\ \mathrm{e}^{\frac{-t}{T_2}}(M_y(0)\,\cos(\omega_0 t) - M_x(0)\,\sin(\omega_0 t)) \end{bmatrix} \qquad (2.17)$$

2.1.2 Time Update Form of Bloch Equation

The Bloch equation in the rotating frame is computed by time updated computation of spin magnetization vector at each spatial location in the x-y plane of the selected slice. For any pulse sequence, the spin precession frequency in the rotating frame of reference is changed by application of external spatial encoding gradients. The effect of RF tipping is simulated as a hard pulse that tips each spatial magnetization vector into the x-y plane. Following application of the hard pulse, strength of the initial magnetization is computed using

$$M_0 = \frac{\rho_0 \gamma^2 \hbar^2}{4kT}(B_0 + \Delta B_0) \qquad (2.18)$$

where ρ_0 is the proton density, γ is the gyromagnetic ratio, $\hbar$ is the Plank's constant, k is the Boltzmann constant, T is the tissue temperature, B_0 is the main magnetic field and ΔB_0 is the static field inhomogeneity. Once the initial magnetization is computed using (2.20), the Bloch equation is solved by stepping forward through time, while approximating the evolution of each magnetization vector using

$$\vec{M}(\vec{r}, t + \Delta t) = Rot_z(\theta_g) \cdot Rot_z(\theta_{iH}) \cdot R_{relax}.R_{RF} \cdot \vec{M}(\vec{r}, t) \qquad (2.19)$$

Here, θ_g and θ_{iH} denote the phase accumulated due to the applied time-varying gradient, and field inhomogeneities respectively. $Rot_z(\theta)$ represents the rotation matrix used to represent rotation of the tipped magnetization vector about z-axis by an angle θ.

$$Rot_z(\theta) = \begin{bmatrix} \cos\theta & \sin\theta & 0 \\ -\sin\theta & \cos\theta & 0 \\ 0 & 0 & 1 \end{bmatrix} \qquad (2.20)$$

In a given time interval Δt, the angle of precession is determined by the sum of angular components contributed by the spin precession frequencies due to the applied gradients $\theta_g = \omega_y \Delta t$ and local field inhomogeneities $\theta_{iH} = \gamma \Delta B \Delta t$ respectively. If the applied gradients are time varying, the phase accumulated in a given time interval Δt is given by

$$\theta_g = \gamma \int\limits_{t}^{t+\Delta t} \left(x G_x(\tau) + y G_y(\tau) \right) d\tau \tag{2.21}$$

The matrix R_{relax} describes the effects of transverse and longitudinal relaxations [8]. The diagonal elements of this matrix indicate the change in signal magnitudes due to relaxations in the x, y and z components respectively.

$$R_{relax} = \begin{bmatrix} e^{\frac{-\Delta t}{T_2(\vec{r})}} & 0 & 0 \\ 0 & e^{\frac{-\Delta t}{T_2(\vec{r})}} & 0 \\ 0 & 0 & 1 - e^{\frac{-\Delta t}{T_1(\vec{r})}} \end{bmatrix} \tag{2.22}$$

If the assumption of a hard pulse is not considered, the RF pulse combines the effect of rotations about z-axis, and flipping about the x or y axis. With the application of a circularly polarized RF pulse: $B_{RF} = B_1(\cos(\omega t) + j \sin(\omega t))$ to tip the magnetization by a flip angle α, the combined effects of rotation and flip are represented using the matrix

$$R_{RF} = Rot_z(\omega_1 t) \cdot Rot_x(\alpha) \cdot Rot_z(-\omega_1 t) \tag{2.23}$$

where $\omega_1 = \gamma B_1$. In the case where the local field differs from B_0 due to field inhomogeneity or chemical shift effects, the resonance frequency ω_{RF} deviates from ω_0 as defined in (1.1). Consequently, the effective precessional frequency becomes

$$\omega_{eff} = \sqrt{(\omega_{RF} - \omega_0)^2 + \omega_1^2} \tag{2.24}$$

A B_1 pulse of duration τ_p tips $\vec{M}$ by an angle α from B_0, resulting in an effective flip angle

$$\alpha_{eff} = \tau_p \sqrt{(\omega_{RF} - \omega_0)^2 + \left(\frac{\alpha}{\tau_p}\right)^2} \tag{2.25}$$

2.2 Working Principle of MR Simulator

MRI head phantom forming input to the simulator, consists of individual spatial maps representative of the tissue relaxation parameters, proton density etc. In specific imaging applications, additional parametric maps may be called for. For example, in Susceptibility Weighted Imaging (SWI), this corresponds to a map of intrinsic susceptibility variations. In PC-MRA, the same parametric maps are used to differentiate vascular regions from background tissue.

Once the required input parametric maps are provided, the simulator solves the Bloch equation using the time-update form, and computes the spin magnetization vector at each spatial position in the transverse (x-y) plane of a given slice [7–10]. By applying a 90° hard pulse along the phase-encode direction, the Free Induction Decay (FID) signals in the *inphase* and *quadrature* coils are obtained as the x and y components of the magnetization vector at each time point. A complex signal is then formed by treating the x-component as the real part, and y-component as the imaginary part. The time update solution is formulated by combining effects of precession and relaxation in each time interval. In matrix form, the time-update form of solution is

$$\vec{M}(t + \Delta t) = \mathbf{A}\vec{M}(t) + \mathbf{B} \tag{2.26}$$

where the matrix parameters $\mathbf{A}$ and $\mathbf{B}$ are determined from $Rot_z(\theta)$ and R_{relax} as

$$\mathbf{A} = \begin{bmatrix} e^{-\Delta t/T_2} & 0 & 0 \\ 0 & e^{-\Delta t/T_2} & 0 \\ 0 & 0 & e^{-\Delta t/T_1} \end{bmatrix} \begin{bmatrix} \cos\theta & \sin\theta & 0 \\ -\sin\theta & \cos\theta & 0 \\ 0 & 0 & 1 \end{bmatrix}$$

and

$$\mathbf{B} = \begin{bmatrix} 0 \\ 0 \\ 1 - e^{-\Delta t/T_1} \end{bmatrix} \tag{2.27}$$

As discussed in Sect. 2.1.2, the precession angle θ is computed using the product of off-resonance frequency $\Delta\omega$ and time-interval Δt. The off-resonance frequency pertaining to a frequency-encoding pulse of constant amplitude G_x is given by

$$\Delta\omega = \frac{\gamma x G x}{2\pi} \tag{2.28}$$

The net magnetization is obtained by summing up the spatial magnetizations from allocations in the transverse plane [11–13]. Application of a phase-encoding gradient pulse of amplitude ΔG_y and duration τ adds a phase φ to the FID signal originating from spatial location $P(x, y)$. Based on the y-location, the encoded phase at $P(x, y)$ is determined as

$$\varphi = \frac{\gamma y \Delta G_y \tau}{2\pi} \tag{2.29}$$

Implementation of the simulator requires specification of the echo and repetition times (*TE* and *TR*) and also the flip angle (α). The sequence type includes information whether the target image is two or three dimensional. For a 2D image, the slice position and orientation is specified. Volumetric acquisition should specify whether the whole FOV is scanned as one thick volume, or as a series of thin slabs.

2.2.1 Imaging Parameters and K-Space Generation

Prior to scanning, all MR imaging sequences require specification of the FOV and
required image resolution. The FOVs are separately specified in the x and y
directions as FOV_x and FOV_y respectively. The image resolution is specified using
the number of voxels in the x and y directions, with the y-direction typically being
the phase-encoding direction. For an axial scan, the phase-encode typically corresponds to the Anterior-Posterior (AP) direction, and the frequency-encoding from
the Left-Ear to Right-Ear (LR) respectively. The sampling intervals in the frequency
and phase-encode directions are

$$\Delta k_x = \frac{1}{FOV_x}$$
$$\Delta k_y = \frac{1}{FOV_y} \tag{2.30}$$

Also, from the Fourier imaging principle described in Sect. 1.3, the k-space
sampling intervals are also related to the frequency and phase-encode pulse
amplitudes using

$$\Delta u = \frac{\gamma G_x \Delta t}{2\pi}$$
$$\Delta v = \frac{\gamma \Delta G_y \tau}{2\pi} \tag{2.31}$$

Noting (2.30) through (2.31), the step-size for phase encoding is obtained as

$$\Delta G_y = \frac{2\pi}{\gamma \tau FOV_y} \tag{2.32}$$

The entire k-space is filled by varying the amplitude of phase-encoding pulse
using integral multiples of ΔG_y. For M phase-encoding steps,

$$G_y = \begin{cases} -\frac{M}{2}\Delta G_y \ to \ \left(\frac{M}{2} - 1\right)\Delta G_y, & for \ M \ even, \\ -\frac{(M+1)}{2}\Delta G_y \ to \ \left(\frac{M+1}{2}\right)\Delta G_y, & for \ M \ odd. \end{cases} \tag{2.33}$$

Each spatial frequency component (k_x, k_y) is represented by an individual data
point $S(\vec{k})$ in the k- space as discussed in Chap. 1. The k-space contains complex
data with the real part denoting the *inphase* (*I*), and the imaginary part, the
quadrature (*Q*) component of magnetization. Each different combination of the
gradient pulses is considered to move the acquisition point in the k-space. Changing
the step-size of phase-encoding pulse amplitudes, fills the k-space line-by-line. The
procedural steps for MR simulation is summarized in the block schematic shown in
Fig. 2.1.

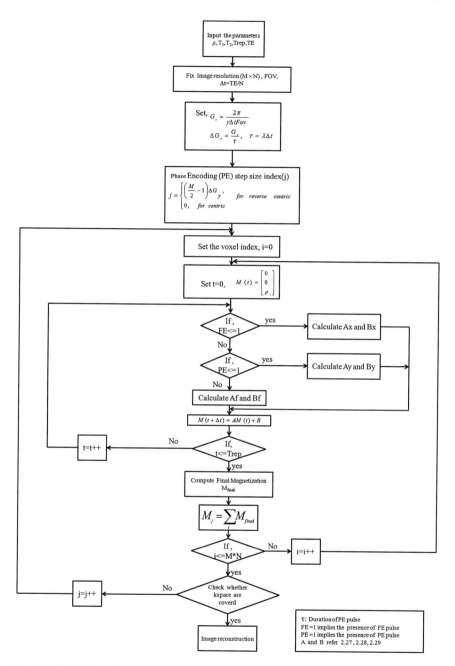

Fig. 2.1 Work flow of MR simulation

2.3 Incorporation of T_2^* Effects in Gradient-Echo Imaging

Intra-voxel field inhomogeneities lead to deviations in the transverse decay rate of spins from the same isochromat. The cumulative effect of all spins within the same voxel result in much faster decay rate of transverse magnetizations [14], than would be predicted by natural atomic and molecular mechanisms. The T_2^* value can be considered as an effective modification of the basic T_2 value of the tissue being imaged. T_2^* effect essentially originates from inhomogeneities in the main magnetic field. The inhomogeneities arise due to intrinsic defects in the magnet itself, or susceptibility-induced field distortions produced by the tissue. Some MR sequences using gradient-echoes acquired using long TE values are relatively more T_2^*-weighted. In MR simulation, the T_2^* effect is mathematically represented by the perturbations of the basic relaxation rate with an additional frequency term generated by the local field inhomogeneity. This is mathematically represented as

$$\frac{1}{T_2*} = \frac{1}{T_2} + \gamma \Delta B_i \qquad (2.34)$$

2.4 Incorporation of Susceptibility Effects

The magnetic susceptibility (χ) denotes the degree of magnetization of a material in response to an applied magnetic field. Since the tissues are diamagnetic and concentrations of paramagnetic agents are relatively small, the susceptibility assumes very small values. Once the susceptibility values are known, it is then easy to find a relation between field changes and susceptibility variations.

$$\Delta B = \chi B_0 \qquad (2.35)$$

Susceptibility effects [15–17] present in brain tissues, lead to off-resonance frequency shifts in *ppm* range. The *ppm* values are appreciably different for venous and arterial structures. The intrinsic susceptibility effects are responsible for phase changes across vessels. Unlike intrinsic susceptibility changes in the *ppm* scale, the bulk susceptibility changes occur at a relatively larger scale. The bulk susceptibility effects mainly occur at air gaps and border regions where the susceptibility values are practically zero on one side of the border. For simulation of intrinsic susceptibility effects, blood vessels are modeled as long cylinders in which the susceptibility values differ inside and outside. Since phase is directly proportional to the local field, it is possible to visualize the field dependence inside and outside of objects embedded in the background. With $\chi_i(\chi_e)$ representing the inside (outside) susceptibility of an object, the $\Delta\chi$ will be equivalent to difference between χ_i and χ_e. The background shift will then depend on the geometry of the structure.

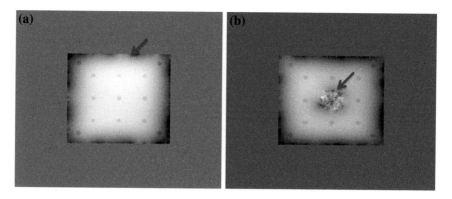

Fig. 2.2 a Simulated phase image without air gap. **b** Simulated phase image with air gap

Consider two analytical phantoms with built in cylindrical structures parallel to the main field (z-direction). The transverse plane is constructed with a finite resolution (256×256), with the cross-section of each cylindrical structure having a radius equivalent to three pixels. The pixels inside the cylinders are assigned a susceptibility value of $\chi_i = 9.45 \times 10^{-6}$, and those outside with $\chi_o = 9 \times 10^{-6}$. A 3D phantom is then constructed by cascading a number of axial cross-sections, large enough to simulate the effect of an infinitely long cylindrical structure. In a slightly different version of the phantom, a central air gap (128×128) is introduced by assigning zero susceptibility in this region. The phase due to spatially varying magnetic field B_0 in both phantoms is calculated from the respective 3D susceptibility maps using [18]

$$\varphi(r) = -\gamma B_0 TE \cdot FT^{-1}\left[FT(\chi(r)) \cdot \left(\frac{1}{3} - \frac{k_z^2}{k_x^2 + k_y^2 + k_z^2}\right)\right] \qquad (2.36)$$

The magnitude images are considered to have uniform intensity in all pixels except zeros in regions containing air gaps. The simulated phase images are shown in Fig. 2.2.

2.4.1   Susceptibility Artifacts

Susceptibility artifacts are seen in the magnitude images as dark regions surrounding borders of interfaces having bulk susceptibility changes [19, 20]. The gradients in susceptibility cause dephasing of spins and frequency shifts of the surrounding tissues, resulting in bright and dark areas with spatial distortion of surrounding anatomy. These artifacts are worst for long echo times in gradient-echo sequences. The susceptibility artifacts can be reduced by performing imaging with low magnetic field strength, smaller voxel size, reduced echo time and increased receiver bandwidth.

SWI uses both phase and magnitude information for the enhancement of venous vasculature. In a typical SWI scan, venous contrast is suppressed in magnitude image due to signal loss in regions with severe field inhomogeneity and peripheral regions with bulk susceptibility changes. SWI processing is performed to enhance the vascular features seen in the magnitude images. The processing, generally involves deriving weights from the high-pass filtered phase image. High-pass filtering is inevitable since it will remove the phase wraps and thereby enhance the high-frequency information. For long echo-times, the magnitude images will have reduced signal intensities due to T_2^* decay. To simulate the effect of susceptibility artifacts at long echo times, a complex image is generated using the phase synthesized from bulk susceptibility changes and uniform magnitude at all tissue locations.

Since vascular information is preserved in the high-frequency phase, SWI processing involves application of a homodyne filter to the phase image. The homodyne filtered phase is given by

$$\theta_{filt} = angle\left(\frac{I}{FFT^{-1}(h \cdot FFT(I))}\right) \qquad (2.37)$$

where θ_{filt} is the high-pass filtered phase, I is the complex image, and h is a 3D transfer function representing the low-pass filter. The transfer function is a Gaussian low-pass filter. If N_i corresponds to the number of voxels in the ith dimension, and σ is a parameter which determines the strength of high-pass filtering, then the filter frequency—response is obtained as

$$H(k_x, k_y, k_z) = \exp\left(-\frac{\left(k_x - \frac{N_x}{2}\right)^2 + \left(k_y - \frac{N_y}{2}\right)^2 + \left(k_z - \frac{N_z}{2}\right)^2}{2\sigma^2}\right) \qquad (2.38)$$

Fig. 2.3 Susceptibility artifact simulated using MATLAB

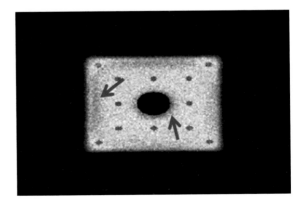

In SWI processing, the magnitude image is point-filtered using weights derived from the high-pass filtered phase. Relative magnitude of the weights is determined from the difference between internal and external fields at locations occupied by venous elements. If the internal field is larger, then the differences are positive resulting in negative phase values. In this case, the weights decrease from 1 to 0 as the phase values increase from 0 to π in the negative direction. Figure (2.3) shows the magnitude image after application of weights. The susceptibility artifacts are seen as dark patches at the border regions, indicated by the arrows.

References

1. Haacke EM, Brown RW, Thompson MR, Venkatesan R (1989) Magnetic resonance imaging: physical principles and sequence design. Wiley, New York
2. Cattin HB, Collewet G (2005) Numerical implementation of the Bloch equations to simulate magnetization dynamics and imaging. J MRI 173
3. Klepaczko A, Szczypinski P, Dwojakowski G, Strzelecki M, Materka A (2014) Computer simulation of magnetic resonance angiography imaging: model description and validation. Plosone 9:e93689
4. Cattin HB, Collewet G, Belaroussi B, Jalmes HS, Odet C (2005) The SIMRI project: a versatile and interactive MRI simulator. J Mag Res 173:97–115
5. Jurczuk K, Kretowskiv BJJ, Eliat PA, Jalmes HS, Wendling JB (2013) Computational modeling of MR flow imaging by the lattice Boltzmann method and Bloch equation. Mag Res Imaging 31:1163–1173
6. Jurczuk K, Kretowski M, Eliat PA, Jalmes HS, Wendling JB (2014) *In Silico* modeling of magnetic resonance flow imaging in complex vascular networks. IEEE Trans Med Imaging 33:11
7. Kwan RKS, Evans AC, Pike GB (1999) MRI simulation-based evaluation of image-processing and classification methods. IEEE Trans Med Imaging 18(11):1085–1097
8. Cao Y, Shen Z, Thomas MA, Chenevert L, Ewing JR (2006) Estimate of vascular permeability and cerebral blood volume using Gd-DTPA contrast enhancement and dynamic T2*-weighted MRI. Magn Reson Imaging 24:288–296
9. Spritzer CE, Pelc NJ, Lee JN, Evans AJ, Sostman HD, Riederer SJ (1999) MRI simulation-based evaluation of image-processing and classification methods. IEEE Trans Med Imaging 18:11
10. Bittoun J, Taquin J, Sauzade M (1984) A computer algorithm for the simulation of any nuclear magnetic resonance (NMR) imaging method. Magn Reson Imaging 2:113–120
11. Duyn J (2013) MR susceptibility imaging. JMR 299:198–207
12. Ruan C. MRI artifacts: mechanism and control. http://ric.uthscsa.edu/personalpages/lancaster/D12_Projects_2003/MRI_Artifacts.pdf
13. Bernstein MA, King KF, Zhou XJ (2004) Hand book of MRI pulse sequences. Elsevier Academic Press, Amsterdam
14. Haacke EM, Mittal S, Wu Z, Neelavalli J, Cheng YC (2009) Susceptibility-weighted imaging: technical aspects and clinical applications, Part 1. AJNR Am J Neuroradiol 30:19–30
15. Cao Z, Oh S, Sica CT, Mc-Garrity JM, Horan T, Luo W, Collins CM (2014) Bloch-based MRI system simulator considering realistic electromagnetic fields for calculation of signal, noise, and specific absorption rate. Magn Reson Med 72:237–247
16. Jochimsen TH, Sch€afer A, Bammer R, Moseley ME (2006) Efficient simulation of magnetic resonance imaging with Bloch–Torrey equations using intra-voxel magnetization gradients. J Magn Reson 180:29–38

17. Gudbjartsson H, Patz S (1995) Simultaneous calculation of flow and diffusion sensitivity in steady-state free precession imaging. Magn Reson Med 34:567–579
18. Marshall I (1999) Simulation of in-plane flow imaging Concepts. Magn Reson 11:379–392
19. Yarnykh VL (2010) Optimal radiofrequency and gradient spoiling for improved accuracy of T1 and B1 measurements using fast steady state techniques. Magn Reson Med 63:1610–1626
20. Neelavalli J, Cheng YC, Jiang J, Haacke EM (2009) Removing background phase variations in susceptibility-weighted imaging using a fast Forward-Field Calculation. J Magn Res Imaging 29:937–948

Chapter 3
Working Principle of PC-MRA with MATLAB Examples

Abstract This chapter summarizes the effects of flow in PC-MRI and introduces methods used to analyze and quantify flow. The chapter begins with a brief introduction to low flip angle GRE sequence and need for velocity encoding. Explanation of phase contrast techniques is provided based on the concept of velocity encoding, followed by individual sections devoted to each specific type of quantitative flow imaging analysis used.

Keywords GRE · Velocity encoding · PCMR techniques · Balanced four point method

As discussed in Chap. 1, Phase contrast techniques derive contrast between flowing blood and stationary tissues by manipulating phase of the magnetization vector. Phase is a measure of how far the magnetization precesses from the time it is tipped into the transverse plane until the time it is detected. Unlike that of spin echoes, gradient echo imaging has no 180° pulse to refocus the effects of differences in resonant frequency due to moving spins. Since the gradients here do not apply refocusing pulse, the magnetic field inhomogeneities are retained in this sequence, thereby introducing a T_2^* weighting. As a result, the phase of moving spins tends to continually change with time upon application of a gradient pulse. Since PC methods are less susceptible to saturation effects, very short TRs (<25 ms) can be used.

A spin echo sequence has a fairly long TR due to the use of 180° RF pulse, and the time needed for longitudinal magnetization to grow back. The use of rather short TE in gradient echo permits TR to be very short. Typical gradient echo imaging sequences have TE of approximately 10–20 ms and this allows TR to be approximately 25 ms. To compensate for the insufficient magnetization recovery due to the short TR, an excitation (flip) angle of less than 90° is commonly used.

© The Author(s) 2016
J. Suresh Paul and S. Gouri Raveendran, *Understanding Phase Contrast MR Angiography*, SpringerBriefs in Electrical and Computer Engineering, DOI 10.1007/978-3-319-25483-8_3

33

3.1 Gradient Echo Imaging

The basic idea of a gradient echo sequence is to obtain the echo only with the help of gradient rephasing, without the need for a 180° RF pulse used in spin echo sequences. The sequence starts with an RF pulse that flips the magnetization vector by an angle $\alpha < 90°$ as shown in Fig. 3.1. An initial negative lobe of the frequency encoding gradient dephases the transverse magnetization (a). The negative lobe turns to a positive lobe of same amplitude (b) that rephases the transverse magnetization and forms a gradient echo at time TE. A gradient pair like this is called a bipolar gradient. Since the slice selecting gradient is not allowed to leave any phase information, the positive slice selecting gradient is also compensated by a corresponding negative lobe.

If a short TR and large flip angle are used, there will be some remnant magnetization left in the transverse plane when the next RF pulse is emitted. This results in artifacts in the final image. To avoid these artifacts, a technique called spoiling is often used. Spoiling is a way to destroy (spoil) the magnetization remaining in the transverse plane after signal sampling, and before the next RF pulse.

Consider a spatial voxel characterized by an unknown spin density distribution. In the absence of gradients, signal loss occurs due to transverse relaxation only. Upon application of a gradient in the x-direction, the local spins within the voxel undergo an additional dephasing [1]. To understand the cause of this dephasing, consider a uniform spin distribution pattern in the voxel centered at x_0. The spin precession at each location 1–5 within a voxel of thickness Δx and centered at x_0 is shown in Fig. 3.2. Since the gradients lead to field perturbations with opposite

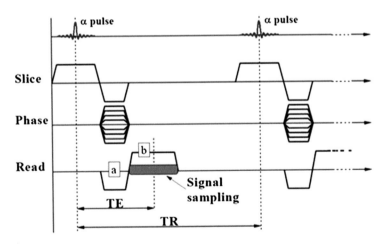

Fig. 3.1 A gradient echo sequence. The uppermost line indicates when the RF pulse is applied. Line two, "slice", shows the slice selecting gradients. The phase encoding gradient is seen in line three, "Phase"; its amplitude is changed as the sequence is repeated. In line four, "read", the gradient lobes form a gradient echo at TE. Signal is collected during the rephasing lobe of the echo

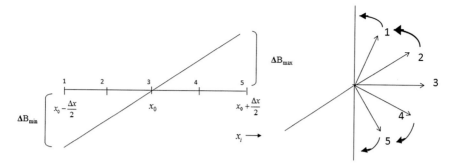

Fig. 3.2 Dephasing of spins due to presence of gradient within a voxel

polarity, the signal components from symmetrically located positions on both sides of x_0 tend to cancel out resulting in a dephasing effect [1].

Neglecting T_2 relaxation, the signal in the absence of gradient will be

$$S(x_0, t) = \int_{-\Delta x/2}^{-\Delta x/2} \rho(x_i)\, e^{-j\omega_i t} dx \tag{3.1}$$

Following application of the negative gradient for a duration τ,

$$S(x_0, t) = \int_{-\Delta x/2}^{+\Delta x/2} \rho(x_i) e^{-j(\omega_i t + \gamma G x_i)} dx \tag{3.2}$$

In this duration, the cumulative phase accumulated will be

$$\varphi_-(t) = -\int_0^\tau dt\, \omega(t)$$
$$= -\gamma \int_0^\tau G_x(t) dt \tag{3.3}$$

3.2 Velocity Encoding

In PC-MRI, the phase effects [2, 3] are sensitized to motion using Velocity Encoding (VENC) gradient prior to readout. Subtraction of two phase images acquired with opposite polarities of VENC allows quantitative assessment of spin velocities [4, 5]. The phase dependency of MR signal to moving spins is derived from the spin precession frequency given by

$$\omega_L(\vec{r}, t) = \gamma B_z(\vec{r}, t) = \gamma B_0 + \gamma \Delta B_0 + \gamma \vec{r}(t) \vec{G}(t) \tag{3.4}$$

where $\vec{G}$ is applied magnetic field gradient, $\vec{r}$ is the spatial location, and ΔB_0 is local field inhomogeneity. The term $\gamma \Delta B_0$ denotes local field change due to static field inhomogeneities and local susceptibility changes. In the rotating frame, the main field contribution to signal frequency can be omitted as discussed in Sect. 1.2. Phase of the precessing magnetization [6] after application of excitation pulse at t_0 and echo time TE is

$$\varphi(\vec{r}, TE) - \varphi(\vec{r}, t_0) = \int_{t_0}^{TE} \omega_L(\vec{r}, t) dt = \gamma \Delta B_0 (TE - t_0) + \gamma \int_{t_0}^{TE} \vec{G}(t) \vec{r}(t) dt \tag{3.5}$$

When expanded as a Taylor series,

$$\varphi(\vec{r}, TE) = \varphi_0 + \sum_{n=0}^{\infty} \gamma \frac{\vec{r}^{(n)}}{n!} \int_{t_0}^{TE} \vec{G}(t)(t - t_0)^n dt \tag{3.6}$$

where $r^{(n)}$ is nth derivative of the time dependent spin position, and φ_n is the corresponding nth order phase [4, 7–11]. Initial signal phase and field inhomogeneities result in additional background phase φ_0. Considering velocity of blood in the range 0.1–1.0 mm/sec, the sampling time Δt to be 1 ms, and spatial resolution to be 0.5 mm, a moving spin traverses 1 μm in one sampling interval. In this case, the time taken to traverse a voxel distance will be 500 ms. If motion of the spin does not change fast with respect to temporal resolution of data acquisition, flow velocities can be approximated to be time invariant during one acquisition [4]. Thus, $\bar{r}(t)$ can be introduced as a first order displacement $\vec{r}(t) = r_0 + \vec{v}(t - t_0)$ with constant velocity $\vec{v}$. With the above approximation, (3.6) can be simplified to

$$\varphi(\vec{r}, TE) = \varphi_0 + \gamma \vec{r}_0 \int_0^{TE} \vec{G}(t) dt + \gamma \vec{v} \int_0^{TE} \vec{G}(t) t dt + \cdots \tag{3.7}$$

Gradient moments M_0 M_1

where φ_0 is an unknown background phase, M_0 (static spins) and M_1 (moving spins) denote zero and first order components describing the influence of magnetic field gradients on the phase of static spins at $\vec{r}_0$, and moving spins with velocity $\vec{v}$. The first gradient moment M_1 determines the velocity induced signal phase for the constant velocity approximation. In this way, appropriate control of the first gradient moment can be used to specifically encode spin flow or motion. Assuming the

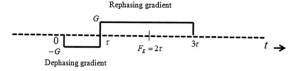

Fig. 3.3 Gradient pulse for echo formation

spins to undergo translational motion with a constant velocity v_x, the phase accumulated during the dephasing lobe of the gradient pulse in Fig. 3.3 is [10]

$$\varphi_-(t) = -\gamma G \int_0^\tau (x_0 + v_x t)dt$$

$$= -\gamma Gx_0\tau - \frac{1}{2}\gamma Gv_x\tau^2 \tag{3.8}$$

In the duration from τ to t of the rephasing lobe, the additional phase accumulated will be

$$\varphi_+(t) = -\gamma Gx_0(t - \tau) - \frac{1}{2}\gamma Gv_x(t^2 - \tau^2) \tag{3.9}$$

Total phase accumulated from the gradient pulse onset is

$$\varphi(t) = \varphi_-(t) + \varphi_+(t) \tag{3.10}$$

$$= -\gamma Gx_0(t - 2\tau) - \frac{1}{2}\gamma Gv_x(t^2 - 2\tau^2) \tag{3.11}$$

Letting $t' = t - 2\tau$, the time origin is shifted to the echo center. The phase expressed using t' can be represented as the sum of a stationary phase term $\varphi_s(t')$ and a motion dependent term $\varphi_v(t')$ given by

$$\varphi_s(t') = -\gamma Gx_0 t' \tag{3.12}$$

and

$$\varphi_v(t') = -\frac{1}{2}\gamma Gv_x(t'^2 + 4\tau t' + 2\tau^2) \tag{3.13}$$

The temporal variation of two components is depicted in Fig. 3.4.

Denoting the first order moments to be $M_1^{(1)}$ and $M_1^{(2)}$, and phase images to be $\varphi^{(1)}$ and $\varphi^{(2)}$ for two bipolar encoded images acquired with leading positive gradient for the first acquisition and leading negative gradient for the second, both phase difference ($\Delta\varphi$) and velocity (v) can be expressed in terms of the change in

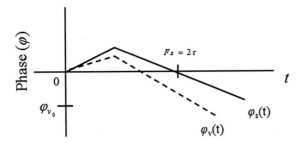

Fig. 3.4 Phase as a function of time for stationary spins (*solid line*) and additional phase of spin moving with constant velocity (*dashed lines*) along direction of G

moments $\Delta M_1 = M_1^{(1)} - M_1^{(2)}$. Neglecting higher order terms, $\Delta\varphi$ can be related to ΔM_1 using

$$\Delta\varphi = \varphi^{(1)} - \varphi^{(2)} = \gamma v \Delta M_1 + \cdots \qquad (3.14)$$

From (3.14), the velocity image is obtained as

$$v = \frac{\Delta\varphi}{|\gamma\Delta M_1|} \qquad (3.15)$$

The sequence diagrams for differential flow measurement are shown in Fig. 3.5. For the design of an actual phase contrast MR measurement, prior knowledge of the order of the maximum velocities is required. For too high velocities, the velocity dependant phase shift exceed $\pm\pi$ and phase aliasing occurs. The Velocity Sensitivity (VENC) is defined as the velocity that produces a phase shift $\Delta\varphi = \pi$ radians. From (3.15), this is equivalent to

$$VENC = \frac{\pi}{\gamma\Delta M_1} \qquad (3.16)$$

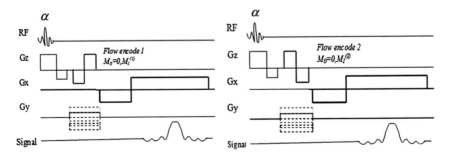

Fig. 3.5 Gradient echo pulse sequence for one directional velocity encoding along the slice direction using bipolar gradients with opposite polarity

The *VENC* can be varied by changing the area of the gradient lobes, thereby making the sequence sensitive to larger flow magnitudes. In the simplest form of a PC sequence, the flow is sensitive to gradient changes only in one direction.

3.3   Effects of Flow on the Image

The effects of motion on spin phase is considered using signal estimated for flowing spins along the read direction. The types of flow considered are plug and laminar flow [12–15]. Plug flow is a constant flow where the velocity is independent of location in the vessel. Laminar flow is a more realistic flow representation, where velocity varies as a function of position. As the mean velocity increases, laminar flow gradually breaks down due to an added random component. In this case, the flow is said to become turbulent.

Introduction of motion causes the spatial coordinates of isochromats to be updated for every single time-step of the pulse sequence. Time dependence of the isochromat spatial coordinates, allows for the introduction of a motion model for all isochromats within the anatomical object. When laminar flow of a homogeneous liquid within a straight tube is simulated, the displacement at each point simulates motion along one axis only, and expressed as a function of time as $x(t) = x_0 + v_x(t)$. Here x_0 is the initial spin position, and v_x is the velocity determined from the laminar flow profile. The velocity profile of laminar flow within the tube as a function of radius r is given as $v(r) = v_{max}(1 - (r^2/R^2))$. Here, v_{max} is the maximum velocity of liquid at the tube center and R denotes its inner radius.

The flow-related artifacts appear in the regions of complex vascular structures as well as with simple straight vessels, resulting in additional difficulties in image analysis. Therefore, understanding MR flow image formation is important for clinical assessment of a vascular disease in the presence of flow induced artifacts. Eddy current and Phase wraps are typical errors affecting flow analysis. Phase changes owing to spatially and temporally varying image gradients can corrupt phase values that encode flow information [16]. Phase wraps occur if the actual blood flow velocity during image acquisition exceeds the VENC value. In PC-MR measurement for too high velocities, the velocity dependent phase shift can exceed $\pm\pi$ resulting in aliasing artifacts. The highest velocity is used to define VENC amplitude, so as to avoid phase wrapping. Other than aliasing, PC velocity images suffer from noise effects that lead to errors in the acquired velocities [17]. Since image phase is affected by pulse sequence timing, field inhomogeneity, radio frequency effects, magnetic field Eddy currents and motion in other directions, reliable methods involving multiple measurements that depend on phase shifts are to be used.

3.4 Phase Contrast Techniques

There are two methods for PC-MRA acquisition. Incoherent technique and Coherent technique. In the incoherent technique, the two projection images acquired one during rapid flow and other during quiescent flow. Rapid flow image shows signal loss from dephasing, while quiescent flow image is less affected. Subtraction of two images yields an angiogram. In incoherent technique two images are acquired. One with flow compensation and other without flow compensation. The subtraction of two images results in an angiogram. Two methods used for subtraction include the Phase Difference (PD) and other Complex difference (CD).

While the flow magnitude is deduced from signal intensity in the final image, the flow direction is described by its associated phase angle. Flowing spins accumulate a phase angle as determined by the phase of the magnetization vector. The ideal signal vector for the two sequences in case of both flow and stationary spins is shown in Fig. 3.6.

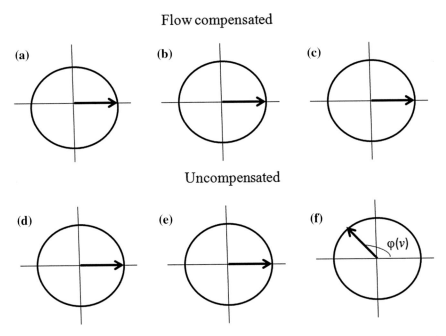

Fig. 3.6 Signal vectors for flowing spins, assuming the vector to have a zero phase angle at $t = 0$ (**a** and **d**). At $t = TE$, the stationary spins (**b** and **e**) yield a totally rephased signal vector both in the flow compensated and the uncompensated sequence. Flowing spins at $t = TE$ (**c** and **f**) are totally rephased in the flow compensated sequence, whereas in the uncompensated sequence, a phase angle is accumulated

Fig. 3.7 Complex difference (CD) vector. When subtracting signal vector elements from a flow compensated sequence (S_f) and an uncompensated sequence (S_u), the flowing spins leave a new vector called CD vector

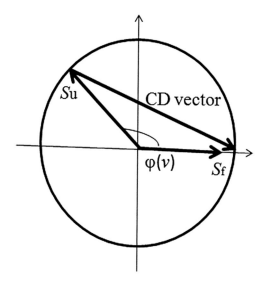

The signal vector matrix from the flow compensated acquisition can be subtracted pixel-by-pixel from the matrix corresponding to the uncompensated acquisition. Pixels containing signal from stationary spins will leave a zero vector [(e)–(b)]. For flowing spins, the result will be a new vector [(f)–(c)], called Complex Difference (CD) vector shown in Fig. 3.7.

Magnitude of this CD vector is dependent on the flow velocity. Low flow velocity yields a small CD vector. As the velocity increases, magnitude of the vector increases until $\varphi(v) = 180°$. For higher flow velocities, the vector magnitude starts to decrease. If the velocity is increased further, a periodic pattern will be observed as shown in Fig. 3.8.

$$|CD| = \sqrt{|S_f|^2 + |S_u|^2 - 2|S_f||S_u|\cos(\varphi)}$$

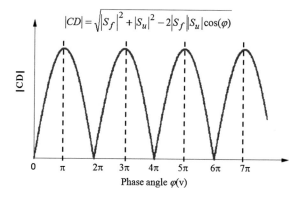

Fig. 3.8 Periodic pattern of the CD magnitude as a function of flow velocity. |CD| may be calculated using the equation in the figure (the cosine theorem). S_f is the signal vector from the flow compensated sequence and S_u is the signal vector from the uncompensated sequence

A Phase Contrast Complex Difference (PC-CD) angiogram is an image showing the magnitude of the CD vector as pixel intensity. This kind of angiogram will ideally have no signal intensity in pixels with stationary spins. Signal intensity in pixels with flowing spins will vary with flow velocity as in Fig. 3.8. The maximum signal intensity is obtained when the flow induced phase angle $\varphi = (2n + 1)\pi$, where $n = 0, 1, 2, \ldots$ Another form of PC angiogram is the called PC Phase Difference (PD) angiogram. In this form of angiographic display, the phase difference φ forms the signal intensity. In a PC-PD angiogram, the signal intensity will be directly proportional to flow velocity as long as the requirement $-180° < \varphi < 180°$ is true.

Encoded velocities can be derived from the data by dividing the pixel intensities in the calculated phase difference images by $\gamma \Delta M$. Only the velocity component along direction of the bipolar gradient, contributes to the phase of MR Signal. It is to be noted that only a single velocity direction can be encoded with an individual measurement. Thus at least four independent measurements with different arrangements of bipolar gradients have to be performed to gain velocity data with isotropic three directional flow sensitivity. The simplest phase contrast method is the Two-Point method [5] that uses two measurements, each made with different magnetic field gradient waveform along one direction to examine motion in that direction. In the Six-Point method, three pairs of measurements are used, each pair sampling the motion along one Cartesian direction.

3.5 Quantitative Flow Image Analysis

3.5.1 Two-Point Method

The two point method uses two measurements, yielding images S_1 and S_2. Contributions to S_1 from static and moving spins are denoted by complex quantities S_s and S_m, having different phase. S_2 is measured with first moment of gradient along direction of v, altered by ΔM_1. The phase of signal from moving spin is then shifted by $\gamma \Delta M_1 v$. With $S_1 = S_s + S_m$ and $S_2 = S_s + S_m e^{j\varphi m}$, the complex difference between two measurements yields

$$\begin{aligned} \Delta S &= S_1 - S_2 \\ &= S_m \left(e^{j\varphi_m} - 1 \right) \\ &= 2jS_m \sin\left(\frac{\varphi_m}{2} \right) \end{aligned} \tag{3.17}$$

From (3.17), it is clearly evident that the complex difference magnitude image eliminates spatial variations in magnitude due to static spins [18, 19]. Hence, the output is independent of any static signal. One of the main limitations is that the magnitude operation discards directional information preserved in the sign of φ_m. In the PD approach, separate reconstruction is performed on each data set, and the resultant phase difference at each point is obtained as

$$\Delta\varphi = \arg\left[\frac{S_2}{S_1}\right] \tag{3.18}$$

The main advantage is retention of directional flow information contained in the sign of $\Delta\varphi$. For a given noise level (σ^2) introduced by the acquisition process, the noise in velocity (σ_v^2) will be large at locations where original signal magnitudes are low. Denoting the signal magnitude by S, the noise variance introduced into the PD image will be

$$\sigma_v^2 = \frac{2\sigma^2}{|\gamma\Delta M_1 S|^2} \tag{3.19}$$

3.5.2   Simple Four Point Method

In Four Point method, a single phase reference is used. The four points correspond to the phase reference φ_0, and the differentially encoded phase differences φ_x, φ_y and φ_z. Due to the use of a common reference, the velocity components are computed by subtracting φ_0 from φ_x, φ_y and φ_z. Pair-wise phase differences from the reference phase yield

$$\hat{v}_x = \frac{\varphi_x - \varphi_0}{\gamma\Delta M_1}$$

$$\hat{v}_y = \frac{\varphi_y - \varphi_0}{\gamma\Delta M_1} \tag{3.20}$$

$$\hat{v}_z = \frac{\varphi_z - \varphi_0}{\gamma\Delta M_1}$$

From (3.20), the four-point PD speed is obtained as

$$\hat{v}_4 = \sqrt{\hat{v}_x^2 + \hat{v}_y^2 + \hat{v}_z^2} \tag{3.21}$$

From the PD noise representation in (3.19), the noise variance introduced into the four-point velocity map takes the form

$$\sigma_4^2 = \frac{\sigma^2}{|\gamma\Delta M_1 S|^2}\left[1 + \frac{(v_x + v_y + v_z)^2}{v^2}\right] \tag{3.22}$$

Similar to the PD approach, noise variance is large at locations where original signal intensities are low in the static image.

3.5.3 Balanced Four Point Method

In this method, the flow encoding first moments are altered in pairs [20–22]. The data acquired with the first sequence is interpreted as a phase reference. For the second sequence, gradient first moment along $\hat{x}$ and $\hat{y}$ directions are altered by ΔM_1. The resultant image can be represented in terms of the reference image as

$$S_2 = S_1 e^{i\varphi_x} e^{i\varphi_y} \tag{3.23}$$

where $\varphi_x = \gamma \Delta M_1 v_x$ and $\varphi_y = \gamma \Delta M_1 v_y$. For third and fourth sequences, the moments along $\vec{x}$ and $\vec{z}$ and $\vec{y}$ and $\vec{z}$ directions are also altered by ΔM_1. In similar lines to (3.26), the images from the third and fourth sequences can be represented as

$$S_3 = S_1 e^{i\varphi_x} e^{i\varphi_z}$$
$$S_4 = S_1 e^{i\varphi_y} e^{i\varphi_z} \tag{3.24}$$

Denoting φ_i to be the phase angle of S_i for $i = 1$ to 4, the phase differences between images from each successive sequence can be expressed as

$$\varphi_2 - \varphi_1 = \varphi_x + \varphi_y$$
$$\varphi_3 - \varphi_1 = \varphi_x + \varphi_z \tag{3.25}$$
$$\varphi_4 - \varphi_1 = \varphi_y + \varphi_z$$

From (3.15), the velocity components along x, y and z directions can be expressed as

$$\hat{v}_x = \frac{-\varphi_1 + \varphi_2 + \varphi_3 - \varphi_4}{2\gamma \Delta M_1}$$
$$\hat{v}_y = \frac{-\varphi_1 + \varphi_2 - \varphi_3 + \varphi_4}{2\gamma \Delta M_1} \tag{3.26}$$
$$\hat{v}_z = \frac{-\varphi_1 - \varphi_2 + \varphi_3 + \varphi_4}{2\gamma \Delta M_1}$$

The term "*balanced*" is used since the variance in the estimated speed is independent of direction. The balanced speed image is obtained as the root sum-of-squares of the component images.

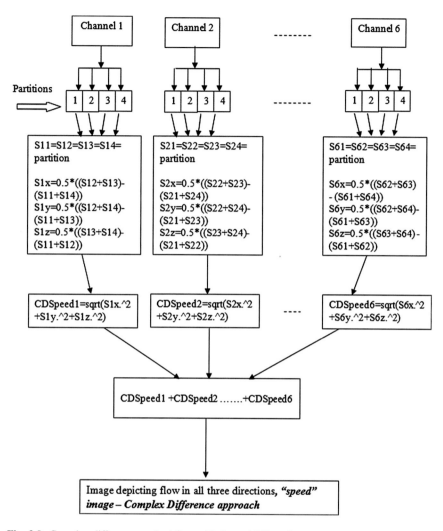

Fig. 3.9 Complex difference method for multi-channel PC angiogram

3.5.4   Processing of Multi-channel PC-MRA

In multi-channel PC-MRA, each channel is acquired as four partitions. Each partition corresponds to the reference and pair-wise encoded data as described in Sect. 3.5.3. In Fig. 3.9, the partitions are represented as S_{cp}, where c represents the channel number, and p represents the partition number. Sample volunteer data were imaged on a Siemens 1.5 T machine with six coils and 64 slices and another with 7

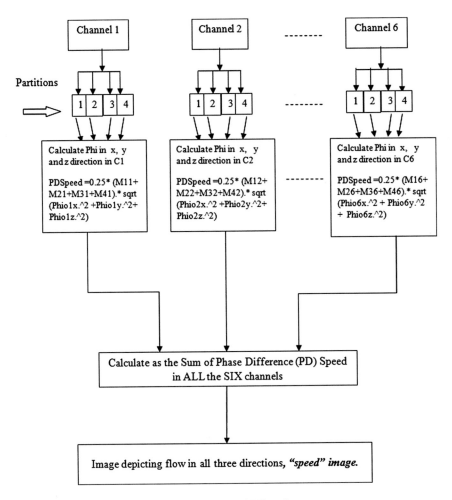

Fig. 3.10 Phase difference method for multichannel PC angiogram

coils 64 slices. The imaging parameters used were $TE = 9$ ms and $TR = 56.70$ ms and flip angle of $15°$. The phase-encoding direction was performed in Anterior-Posterior direction with a VENC value 10 cm/s. The data acquired for each coil consisted of 256 slices. These were then separated into four partitions $S11$, $S12$, $S13$ *and* $S14$ each consisting of 64 slices. The process was repeated for all the channels. CD and PD methods were applied to partitions in each channel using the steps outlined in Figs. 3.9 and 3.10. The combined speed image is computed as shown in block 5 of Fig. 3.9. Sample MIP image are shown in Figs. 3.11 and 3.12. MATLAB codes for computation of CD and PD image are included in Appendix A-1.

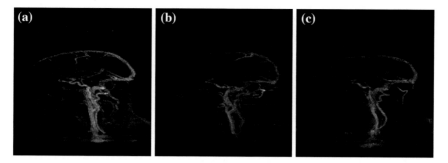

Fig. 3.11 CD speed image computed for three different data sets **a** data-set 1 with 6 channels, **b** data-set 2 with 6 channels, **c** data-set 3 with 7 channels

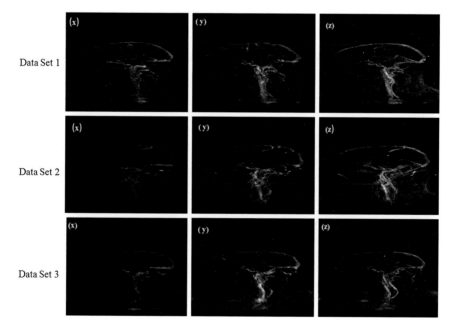

Fig. 3.12 CD speed images in x, y and z directions for three different volunteer data-set

References

1. Haacke EM, Brown RW, Thompson MR, Venkatesan R (1989) Magnetic resonance imaging: physical principles and sequence design. Wiley, New Jersey
2. Dumoulin CL (1995) Phase contrast MR angiography techniques. Magn Reson Imaging 3 (3):399–411
3. Dumoulin CL, Souza SP, Walker MF, Wagle W (1989) Three dimensional phase-contrast angiograph. Magn Reson Med 9:139–149
4. Markl M (2005) Velocity encoding and flow imaging. University Hospital Freiburg, Dept of Diagnostic Radiology, Germany

5. Pelc NJ, Bernstein MA, Shimakawa A, Glover GH (1991) Encoding strategies for three-direction phase contrast MR imaging of flow. Magn Reson Imaging 1(4):405–413

6. Hahn EL (1960) Detection of sea water motion by nuclear precession. J Geophys Res 65 (2):776–777

7. Peeters JM, Bos C, Bakker CJ (2005) Analysis and correction of gradient nonlinearity and B_0 inhomogeneity related scaling errors in two-dimensional phase contrast flow measurements. Magn Reson Med 53(1):126–133

8. Bernstein MA, Shimakawa A, Pelc NJ (1992) Minimizing TE in moment-nulled or flow encoded two- and three-dimensional gradient echo imaging. J Mag Resn Imaging 2:583–588

9. Axel L, Morton D (1987) MR flow imaging by velocity-compensated/uncompensated difference images. J Comput Assist Tomogr 1(1):31–34

10. Haacke EM, Lenz GW (1987) Improving MR image quality in the presence of motion by using rephasing gradients. Am J Roentgenol 148:1251

11. Axel L (1984) Blood flow effects in magnetic resonance imaging. AJR Am J Roentgenol 143 (6):1157–1166

12. O'Donnell M (1985) NMR blood flow imaging using multiecho, phase contrast sequences. Med Phys 12(1):59–64

13. Constantinesco A, Mallet JJ, Bonmartin A, Lallot C, Briguet A (1984) Spatial or flow velocity phase encoding gradients in NMR imaging. Magn Reson Imaging 2(4):335–340

14. Pelc NJ, Sommer FG, LiK C, Brosnan TJ, Herfkens RJ, Enzmann DR (1994) Quantitative magnetic resonance flow imaging. Magn Reson Med 10(3):125–147

15. Korosec FR, Mistretta CA (1998) MR angiography: basic principles and theory. Magn Reson Imaging 6(2):223–256

16. Ahn SJ (2008) Geometric fitting of parametric curves and surfaces. J Inf Process Syst 4:153–158

17. Spritzer CE, Pelc NJ, Lee JN, Evans AJ, Sostman HD, Riederer SJ (1990) Rapid MR imaging of blood flow with a phase sensitive, limited-flip-angle, gradient recalled pulse sequence: preliminary experience. Radiology 176:255–262

18. Bernstein MA, Grgic M, Brosnan TJ, Pelc NJ (1994) Reconstructions of phase contrast, phased array multicoil data. MRM 32:330–334

19. Bernstein MA, Ikezaki Y (1991) Comparison of phase-contrast and complex difference processing in phase contrast MR angiography. J Magn Reson Imaging 1:725–729

20. Nishimura DG, Macovski A, Pauly JM (1987) MR angiography by selective inversion recovery. Magn Reson Med 4:193–202

21. Bryant DJ, Payne JA, Firmin DN, Longmore DB (1984) Measurement of flow with NMR imaging using gradient pulse and phase difference technique. J Comput Assist Tomogr 8:588–593

22. Nayler GL, Firmin DN, Longmore DB (1986) Blood flow imaging by cine magnetic resonance. J Comput Assist Tomogr 10(5):715–722

Chapter 4
Numerical Simulation of PC-MRA

Abstract Quantitative evaluation of image processing algorithms for angiography images can only be approached using synthetic images, where the true geometry of vessels is known. This requires both simulation of flow and evolution of magnetization due to flow. The chapter begins with introduction of flow phantoms presented in a computational perspective. Since the main goal is to understand the evolution of magnetization in PC-MRA, and also because computational models for flow simulation of viscous fluids such as blood are well established, we have excluded their description from the contents of this chapter. The rest of this chapter mainly focuses on the two main methods used for estimation of flow induced magnetization.

Keywords Phantom model · Magnetization transport · Particle trajectory

Flow-induced disturbances can often lead to difficulties in vascular image interpretation. Therefore, assessment of flow in clinical imaging requires prior information on how the flow induced changes affect local field variations. The first section of this chapter presents simulation of flow induced magnetization changes. With flow induced magnetization, local magnetic field is spatially and temporally varying; making the Bloch equation highly non-linear [1, 2]. The update of magnetization at each time step is estimated using the conditions for transport of magnetized fluid particles across a volume element.

From a clinical perspective, modeling abnormal flow conditions such as septal defects in the heart, or streamline patterns in the aorta; require a model to relate the local pressure fields and velocity gradients. This is generally accomplished using the discrete form of Navier-Stokes equation, numerically solved using Finite Element Methods (FEM) [3]. Simulating each stage of the circulatory system, including surges of flow in the heart and stationary flow in the capillary vessels, calls for a range of mathematical models that best describe the appropriate fluid mechanical conditions. These models, ranging from lumped parameter, 1D wave propagation, and 3D numerical methods; can all be used with effect to describe the velocity and pressure fields [4].

© The Author(s) 2016   49
J. Suresh Paul and S. Gouri Raveendran, *Understanding Phase Contrast MR Angiography*, SpringerBriefs in Electrical and Computer Engineering, DOI 10.1007/978-3-319-25483-8_4

One-dimensional method assumes that blood flow velocity along the vessel axis is much greater than flow velocity perpendicular to the vessel axis. In the 1D equations of blood flow, velocity and pressure are averaged over the cross-section of the vessel; resulting in a system of nonlinear partial differential equations in a single spatial variable and time.

For deriving the flow field, blood is considered as incompressible liquid with constant density, with an assumption that the adiabatic movement of liquid is controlled by the Navier-Stokes equation and equation of continuity [5–7]. The solution of Navier-Stokes equation [3, 5] can be applied to find the velocity profile at each cross-section of the vascular tree for input to the MRA simulator. In unsteady flow, the velocity profile at each cross-section is updated at every time-step of flow simulation. The physical domain (Ω) consists of a fluid (vessel) part (Ωf) and a solid part (Ωs) representing background. The interface between the two will be identified as ($\partial\Omega$). Application of mass and momentum conservations at this interface yields

$$\frac{\partial \rho}{\partial t} + \nabla \cdot (\rho \vec{u}) = 0$$

$$\rho \left(\frac{\partial \vec{u}}{\partial t} + \vec{u} \cdot \nabla \vec{u} \right) = -\nabla p + \rho v \nabla^2 \vec{u} + p\vec{f}$$

(4.1)

where $\bar{u}$ is the velocity of fluid with $\nabla \cdot u = 0$, ρ is the mass density, p is the pressure and $\bar{f}$ is a forcing term used to represent the impenetrability of complex shaped solid vessel walls, i.e., the no-slip condition.

After computation of flow fields, the signal generation in PC-MRA is simulated using either the concept of magnetization transport, or particle trajectory tracing. In the former approach, the local magnetization applied as input to the Bloch equation is computed by modeling blood flow in Eulerian (stationary) coordinates using Boltzmann transport of magnetized fluid particles. In the particle tracing approach, the particle velocity u at a given time point is computed along each trajectory, and resolved into the x, y, and z components. The velocity vectors are then input to the Bloch equation, with components of magnetization contributed by the particle projected along each of the three orthogonal directions. Both approaches are discussed in detail in Sects. 4.2 and 4.4. Prior to this, a brief description of flow phantoms that serve as prototype geometric structures for flow field computation is provided.

4.1 Flow Phantom Model

PC-MRA images are simulated using flow phantoms in which 2D and 3D vessel structures are artificially created with characteristics similar to real blood vessels [8, 9]. A typical vascular tree structure can be modeled as shown in Fig. 4.1. In a tree

Fig. 4.1 Vascular tree
structure

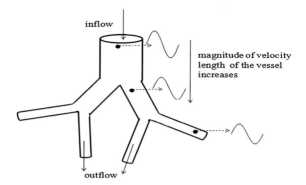

structure, the amplitude of flow waveform decreases with increase in distance from
the source of pressure.

The spatial and temporal flow behavior results from the topology, geometry, and
physiology of the vascular structures. Creation of computational models for MR
flow imaging, helps in understanding this relationship between vascular geometry
and spatio-temporal dynamics of blood flow.

For characterization of vascular geometry, the centerline of a vascular structure
is expressed in parametric form as $(x(s), y(s), z(s))$, with $0 \leq s \leq 1$. To delineate the
Ωf and Ωs part, it is required to determine whether or not the minimal distance from
every point (X, Y, Z) to the centerline is smaller than the radius of the desired
cylindrical vascular structure [10, 11]. A masking function is then generated based
on this minimum distance. The smallest distance approach is applied to centers of
the cells in the 'basic' strategy, and to all eight nodes of the 3D cell in the 'inner'
and 'outer' strategies to be discussed in Sect. 4.1.1. Distance to the centerline is
computed using

$$D = \sqrt{(X - x(s))^2 + (Y - y(s))^2 + (Z - z(s))^2} \qquad (4.2)$$

For every (X, Y, Z), D is a function of the parameter s only. Global minimum of the
distance function is obtained from the first order condition $D' = 2\mathbf{d} \cdot \mathbf{d}' = 0$, where
the prime indicates differentiation with respect to s. This specifies that for the
minimum distance, vector $\mathbf{d}$ is perpendicular to the tangential vector $\mathbf{d}'$ at the
centerline. For a parametric curve, the extremum occurs when s satisfies

$$(X - x(s))x'(s) + (Y - y(s))y'(s) + (Z - z(s))z'(s) = 0 \qquad (4.3)$$

Equation (4.3) can be solved numerically to obtain the local and global extrema for
$0 \leq s \leq 1$. Modeling the centerline as a planar curve, it is possible to set $y' = 0$ and x
$(s) = 4s$, so that

(a)   **(b)**

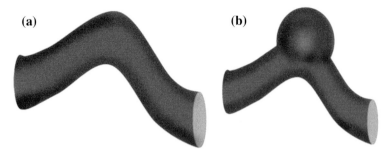

Fig. 4.2 Three dimensional shapes with a circular *center-line*. **a** Curved vessel. **b** Aneurysm model

$$-L_x^2 s + (Z - z(s))z'(s) = -XL_x \qquad (4.4)$$

A sinusoidal centerline is simulated by choosing $y(s) = L_y/2$ and $z(s) = L_z/2 + C \sin [2\pi(s - 1/4)]$, where C is a curve parameter. In order to specify the 'type', i.e., solid or fluid, it is required to determine the parameter-value s at which the global minimum of (4.3) is attained. Numerically, this can be implemented in two stages: first, the distance function is coarsely sampled in $2k$ steps to obtain a 'candidate' interval. Secondly, this interval is refined to obtain the global minimum using simple bisection. Once the optimal s is determined at a given (X, Y, Z), the smallest distance d is computed for this optimal s. If this smallest distance is such that $d \leq R$, then (X, Y, Z) is of type 'fluid', or otherwise. Two basic geometries motivated by medical application are those of curved vessels and model aneurysms [11]. The curved vessel is a cylindrical tube with a sinusoidal centerline. The model aneurysm is the extension of this curved vessel obtained by merging with a sphere as shown in Fig. 4.2.

4.1.1 Masking Function

The masking function technique is a simple and fast way to indicate the location of an object. If the center of the grid cell is of type '*background*' or '*vessel*', then that entire grid cell is taken to be of that type. Based on this rule, Fig. 4.3 illustrates the case where a few cells tend to be inside a vessel, and others in the background.

In three-dimensional domains, this problem is combined in a rectangular block of size $L_x \times L_y \times L_z$ that is large enough to contain the vascular area of interest. For a given number $N_{x,y,z}$ of grid cells, a uniform cartesian grid with mesh-spacings $h_{x,y,z} = L_{x,y,z}/N_{x,y,z}$ is defined for each coordinate direction. If the center of a cell corresponds to background (vessel), then the whole cell is treated to be background (vessel) with the masking function set to $H = 1(0)$.

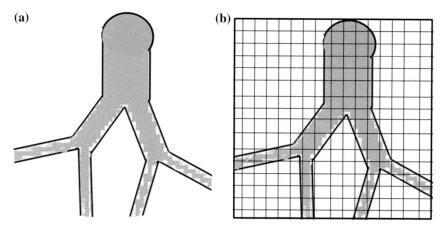

Fig. 4.3 **a** Sketch of flow domain in *grey shaded* area. **b** Flow domain embedded in cartesian grid

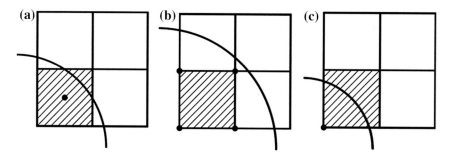

Fig. 4.4 Strategies to define a masking function for a 2D grid cell in a Cartesian grid. Vessel cells are shown hatched. **a** The property at the center of the cell defines that of the whole cell. **b** Cell is classified as vessel if all its corners are inside the vessel domain. **c** Cell is classified as vessel if at least one of its corners belongs to the vessel domain

A second approach uses corner-nodes of the grid cell for deriving the masking function. Applying this to 3D cells, a grid cell is classified as vessel, if all eight of its nodes belong to the vessel class. This is illustrated in Fig. 4.4b, for the 2D case in which the vascular cell is shown hatched. Another strategy assigns the value '*vessel*' for the whole cell, if at least one of its corners is in the vessel part of the domain as shown in Fig. 4.4c.

Masking strategies define the masking function in the middle of a grid cell, i.e., in Hp points of a staggered grid. For a given Hp, Hu and Hv can be extracted for the staggered grid as shown in Fig. 4.5. The value of masking function assigned to a cell face is treated as the maximum of the Hp values of the neighboring grid cells. The mechanism of extracting staggered masking values at the grid faces is illustrated in Fig. 4.5.

Fig. 4.5 Illustration of staggered masking value at grid faces

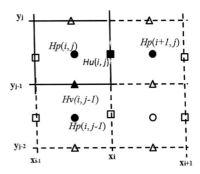

The process of defining masking functions at staggered locations indicated by squares and triangles, is based on the values in the middle of the cell (represented by circles). The maximum of two Hp values of neighboring cells sharing a common face is assigned to the common face as the staggered masking function Hu or Hv. This reflects the condition that if a background and a vessel cell come together in a face, then the face is taken to be solid as well. For a given neighborhood, the masking functions of the staggered grid are computed using

$$Hu(i, j) = \max(Hp(i, j), \ Hp(i + 1, j))$$

and   (4.5)

$$Hv(i, j - 1) = \max(Hp(i, j), \ Hp(i, j - 1)).$$

4.2 Simulation of Magnetization Transport

Modeling of flow requires information about MRI sequence events such as excitation or spatial encoding which occur in time ranges varying from a few *ms* to a few seconds. Magnetized fluid particles are transported during and between those MRI events due to blood flow, and are subjected to constantly changing magnetic conditions. The conditions under which fluid particles present themselves in the MR image are dependent on the characteristics of flow pattern. Since MRI processes are modeled using the discrete time solution of Bloch equation, the resulting analytical transformations of local magnetization due to flow are used to track the signal changes in MRI [12, 13].

During imaging simulation, the local magnetizations are modified according to the MR phenomenon and flow in parallel. The flow influence is modeled using magnetization transport algorithm implemented in Eulerian coordinates (i.e. stationary frame) [14], while the MR processes are modeled by the discrete time solution of the Bloch equation in the rotating frame as discussed in Sect. 2.1.2 [15]. The fluid flow is modeled using Boltzmann transport model of magnetized fluid particles. Boltzmann transport equations are numerically solved using the Lattice

Boltzmann method (LBM) [16–19]. It is based on the evolution of statistical fluid distribution on a lattice/grid (inspired by kinetic theory). The fluid characteristics evolve on the grid nodes, according to rules at the mesoscopic level and concomitantly satisfying the mass and momentum conservation laws at the macroscopic level. The main advantage of LBM when used for vascular flow modeling is its simplicity, and adaptability to model complete vascular networks with complex boundary conditions [20]. A second benefit arises from the low computational load and ease of parallelization [21, 22].

4.2.1   Lattice Boltzmann Method (LBM)

The origin of LBM [6, 16, 23] was from Ludwig Boltzmann's kinetic theory of gases. The basic principle is that the fluids or gases can be imagined as a group of small particles in random motion. In such cases, the momentum and energy is achieved through particle streaming; as modeled by the Boltzmann transport equation

$$\frac{\partial f}{\partial t} + \vec{u} \cdot \nabla f = \Omega \tag{4.6}$$

where $f(r, t)$ is the particle distribution function, u is the particle velocity and Ω is the collision operator. In LBM, the particles are confined to the nodes of a lattice. For 2D model, the particle streams are considered in nine directions, with one staying at rest. The velocities along particle stream directions are referred to as microscopic velocities denoted by e_i, for $i = 0, 1, \ldots, 8$. A 2D model with nine streaming directions is known as the D2Q9 model. The discrete probability distribution function $f_i(r, e_i, t)$ or $f_i(r, t)$ describes the probability of streaming in a particular direction, associated with each particle on the lattice. The geometrical structure of a D2Q9 lattice is depicted in Fig. 4.6. The macroscopic fluid density is determined using

Fig. 4.6 D2Q9 lattice structure

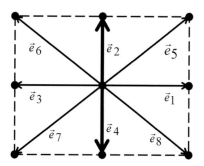

$$\rho(r,t) = \sum_{i=0}^{8} f_i(r,t) \tag{4.7}$$

The macroscopic velocity $u(r, t)$ is computed as an average of the microscopic velocities weighted by the distribution function.

$$u(r,t) = \frac{1}{\rho} \sum_{i=0}^{8} c f_i e_i \tag{4.8}$$

The distribution function at each lattice point is updated using streaming and collision, mathematically represented by

$$\underbrace{f_i(\vec{x} + c\vec{e}_i\Delta t, t + \Delta t) - f_i(\vec{x},t)}_{\text{Streaming}} = -\underbrace{\frac{[f_i(\vec{x},t) - f_i^{eq}(\vec{x},t)]}{\tau}}_{\text{Collision}} \tag{4.9}$$

This equation holds for lattice points within the fluid domain, but not for the domain boundaries. At domain boundaries, the boundary conditions compensate for the insufficient number of distribution functions. Because of this, streaming and collision are computed separately. Figure 4.7 show streaming in interior nodes. In the streaming step, the distribution functions are translated to neighboring sites based on the respective discrete velocity direction.

In collision term of (4.9), $f_i^{eq}(r,t)$ is an equilibrium distribution and τ is a relaxation time to reach equilibrium. The collision process on a D2Q9 lattice is illustrated in Fig. 4.8. For simulating flows of single phase, Bhatnagar–Gross–Krook (BGK) [24, 25] collision is used whose equilibrium distribution is given by

$$f_i^{eq}(r,t) = \omega_i \rho + \rho s_i(u(r,t)) \tag{4.10}$$

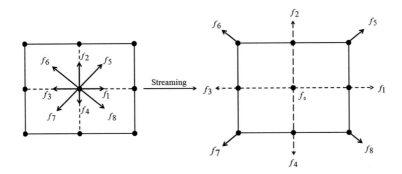

Fig. 4.7 Streaming of interior nodes of D2Q9 lattice

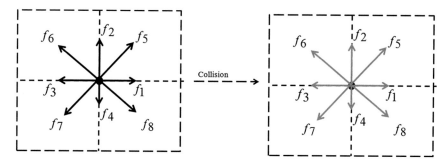

Fig. 4.8 Collision process in a D2Q9 lattice

where $s_i(u)$ and the weights ω_i are defined as

$$s_i(u) = \omega_i \left[3\frac{e_i \cdot u}{c} + \frac{9}{2}\frac{(e_i \cdot u)^2}{c^2} - \frac{3}{2}\frac{u \cdot u}{c^2} \right]$$

and

$$\omega_i = \begin{cases} \frac{4}{9} & i = 0 \\ \frac{1}{9} & i = 1, 2, 3, 4 \\ \frac{1}{36} & i = 5, 6, 7, 8 \end{cases}$$

$c = \Delta x/\Delta t$ denotes the lattice speed. Though the local density ρ and u are conserved, the distribution function changes according to the local-Maxwellian rule [26]. According to this, the pressure p and density ρ for an isothermal state are related using

$$p = c_s^2 \rho \qquad (4.11)$$

For use in Navier-Stokes equation, the fluid kinematic viscosity in the D2Q9 model and relaxation time are related using

$$\nu = c_s^2\left(\tau - \frac{1}{2}\right) \qquad (4.12)$$

Equation (4.12) provides a straightforward method for adjusting the fluid viscosity in the model. It is obvious that the condition $\tau \geq 0.5$ is required to ensure a positive viscosity. The limiting cases of τ tending to 0.5 and ∞ correspond to representations of inviscid flow and Stokes (creeping) flow respectively. While the latter poses no difficulty to the model, the former limit leads to unstable solutions due to insufficient lattice resolution. This is due to the fact that velocity gradients can become very large, and the model cannot dissipate the energy due to the low viscosity. The steps for computing u using LBM are summarized below.

Step 1   Initialize ρ, u, f_i and f_i^{eq}
Step 2   Move f_i to f_i^* in the direction of velocity (streaming)
Step 3   Compute ρ and u from f_i^* using (4.7) and (4.8)
Step 4   Compute f_i^{eq} from (4.10)
Step 6   Calculate updated distribution function $f_i = f_i^* - \frac{1}{\tau}\left(f_i^* - f_i^{eq}\right)$ using
         (4.9)—(collision)
Step 7   Repeat steps 2 to 5

During Step 2 and Step 6, the boundary nodes require unique choice of distribution functions, to satisfy the macroscopic boundary conditions [18].

4.3   Simulation of MRI Signal Generation Using LBM and Bloch Equation

The imaged area is divided into cubic elements with one LBM grid node situated at the center of each cubic element [12, 13]. This guarantees that each cubic element contains information about the flow of fluid filling it. The extended Bloch equation is used to model the behavior of spins in a cubic element at each time point. The Bloch equation with time-varying spatially dependent excitation is

$$\frac{dM(r,t)}{dt} = \gamma(M(r,t) \times B(r,t)) - \frac{M_x(r,t)\hat{i} + M_y(r,t)\hat{j}}{T_2(r)} - \frac{M_z(r,t) - M_0(r)\hat{k}}{T_1(r)}$$

$$(4.13)$$

where M_0 is equilibrium magnetization based on proton density and r is the position vector of the cubic element. The excitation input $B(r, t)$ is given by

$$B(r,t) = B_1^{RE}(r,t)\hat{i} + B_1^{IM}(r,t)\hat{j} + [B_0 + \Delta B(r) + r \cdot G(t)]\hat{k} \qquad (4.14)$$

where $B_1^{RE}\hat{i} + B_1^{IM}\hat{j}$ is the circularly polarized RF pulse in rotating frame, B_0 is the main magnetic field, ΔB is the local magnetic field inhomogeneities, and G is the applied magnetic field gradient at position r. When MRI model is combined with fluid flow, the whole MR imaging process is divided into sufficiently small time-steps. After each time step Δt, the local magnetizations of all cubic elements are evaluated; taking into account both the influence of flow and the MR phenomena. The updated magnetization $M(r, t + \Delta t)$ is calculated considering the flow influence ΔM_{FLOW} and the imaging process A_{MRI}. The recursive estimation is obtained as (see 2.19)

$$M(r, t + \Delta t) = A_{MRI}(r, \Delta t)[M(r, t) + \Delta M_{FLOW}(r, \Delta t)] \qquad (4.15)$$

where ΔM_{FLOW} represents the net flow in cubic element and A_{MRI} represents the magnetic resonance influences during the time period Δt on the magnetization $M(r, t)$ of the cubic element at position r. The details of how ΔM_{FLOW} can be calculated, is included in Sect. 4.3.1. In matrix form, this is similar to (2.19) and given by

$$A_{MRI}(r, \Delta t) = E_{RELAX}(r, \Delta t)R_z(\theta_g)R_z(\theta_{iH})R_{RF}(r, \Delta t) \qquad (4.16)$$

where E_{RELAX} represents the relaxation phenomena and is same as that in (2.22). The angles θ_g and θ_{iH} are similar to those discussed in Sect. 2.1.2.

4.3.1   Integration of LBM and Blochequation Simulation

The modeling environment consists of different steps like geometry specification, geometry distortion, grid generation, flow simulation, and image generation. The magnetization transport algorithm can be combined with the modeling environment for PC-MRA simulation. The flow influence is estimated by tracing the propagation of magnetization through successive elements. For each cubic element at position r, the magnetization changes in a time-step Δt is [27]

$$\Delta M_{FLOW}(r, \Delta t) = \Delta M_{IN}(r, \Delta t) - \Delta M_{OUT}(r, \Delta t) \qquad (4.17)$$

where ΔM_{IN} and ΔM_{OUT} denote the inflow and outflow magnetizations respectively. The inflow and outflow magnetization values for each cubic element are calculated based on the velocity $u = u_x \hat{i} + u_y \hat{j} + u_z \hat{k}$ obtained from the flow model component. While the fraction of inflow magnetization of a cubic element is calculated based on its flow properties and magnetizations of neighboring elements, the fraction of outflow magnetization are calculated from its flow properties and local magnetization. The fraction of magnetization leaving the cube element is represented by the shaded rectangle in Fig. 4.9a.

For the 2D case, the fractions of inflow and outflow magnetizations are computed using

$$\begin{aligned}
\Delta M_{IN}(r, \Delta t) = \; & M\left(\left(r - \Delta d \frac{u_x(r)}{|u_x(r)|}\hat{i}, t\right)\right) |u_x(r)|(1 - |u_y(r)|) \\
& + M\left(r - \Delta d \frac{u_x(r)}{|u_x(r)|}\hat{i} - \Delta d \frac{u_y(r)}{|u_y(r)|}\hat{j}, t\right) |u_x(r)||u_y(r)| \qquad (4.18) \\
& + M\left(r - \Delta d \frac{u_y(r)}{|u_y(r)|}\hat{j}, t\right) (1 - |u_x(r)|)|u_y(r)|
\end{aligned}$$

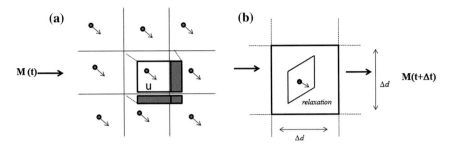

Fig. 4.9 Inclusion of flow effects into the evolution of magnetization. Magnetization in each cubic element is modified based on flow, taking into account the magnetization transfer between neighboring nodes. In each cubic element, MR signal generation is modeled by assigning the value of local magnetization in (4.17) to the Bloch equations. **a** Magnetization transport between neighboring nodes. **b** Magnetization evolution at each node

and

$$\Delta M_{OUT}(r,t) = M(r,t)\big[|u_x(r)|(1 - |u_y(r)|) + |u_x(r)||u_y(r)| + (1 - |u_x(r)|)||u_y(r)|\big]$$

$$(4.19)$$

In the first step of geometry specification, a virtual object which defines regions occupied by different tissue types is selected. In the second step of geometry distortion, the distortions in vascular geometries are introduced by adding and removing spaces bounded by the virtual object. In the grid generation step, the object is divided into cubic elements with a fixed volume. In each cubic element, the influence of MRI events for each successive time step is expressed as in (4.15). In Step 4 representing flow simulation, the velocity maps are generated by performing alternative collision and propagation steps, until the global absolute difference of the velocity fields between successive iterations is smaller than a preset tolerance. A work flow of PC-MRA flow simulation using magnetization transport is depicted in Fig. 4.10a. In the final step, MR signal from the imaged area is computed using

$$S(t) = \sum_{r_0 \in C} M_x(r_0, t)\hat{i} + \sum_{r_0 \in C} M_y(r_0, t)\hat{j} \qquad (4.20)$$

where C is the collection of all cubic elements of the imaged area. The steps in signal generation are summarized in Fig. 4.10b.

The steps describe the sequential procedure employed for computation of flow influence, followed by those required for calculation of MR influence. Integrating all the five steps, a series of images are generated for each time-step. The signal is sampled during an acquisition period and saved in the k-space matrix. Each subsequent excitation is performed with a different phase encoding step, and the acquired signals fill the successive matrix rows. The MR image is created by

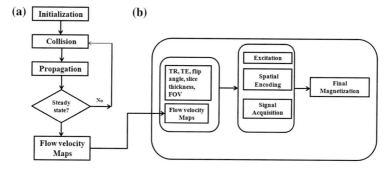

Fig. 4.10 Work flow of flow simulation. **a** Steps for generation of velocity map. **b** Magnetization signal generation

application of the fast Fourier transform to the fully filled matrix. The time-step for magnetization update is equal to the time-step for flow modeling. As per requirement, the time-steps can be changed during various stages of imaging: i.e., shorter during a slice selection and longer after a signal acquisition up to the time point of next excitation. This time-step cannot be longer than the shortest time needed by all the fluid to pass from one grid node to another. It is possible that magnetization fractions can go more than one grid node in a single time step. A shorter time-step enables the time resolution to be increased, leading to an excessive increase of the simulation time. The model usually applies a time-step of the order of tens to hundreds of microseconds and a streaming distance of the order of tens to hundreds of micrometers.

4.4 MRA Simulation Using Particle Trajectory Models

Input to MRA simulator is a model consisting of particle trajectories of a flow field, with particles assigned for each trajectory at the inlet cross-section [10, 28]. The particle is assumed to have spherical shape, and described by its density and diameter. The trajectories are defined based on particle coordinates at each flow simulation time-step Δt. Every particle corresponds to a portion of the fluid, such that the sum of all these portions forms the total volume of blood flowing through the vessels.

A trajectory is defined as an imaginary curve in the flow field, so that the tangent to the curve at any point represents the direction of the instantaneous velocity at that point. The shape of trajectories changes from instant to instant in an unsteady flow, where the velocity vector changes with time. Cross-sections perpendicular to the trajectories are partitioned into discrete nodes using Voronoi partitioning. Figure 4.11 illustrates modeling of flow using trajectories.

The partitions divide the cross section into discrete points, with each point corresponding to an individual trajectory. The micro neighborhood is constructed

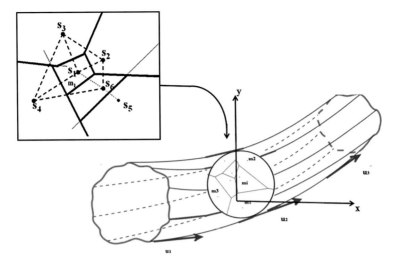

Fig. 4.11 Modeling of flow using trajectories

based on the concept of Voronoi decomposition, where each spatial unit is allotted
its region of influence in terms of an intersecting half plane. A micro neighborhood
around each spatial unit is generated based on the concept of Voronoi polygons.
The micro neighborhood represents the region of influence of a spatial unit over
another spatial unit for any given attribute. For generating a micro neighborhood,
let S = ($s1$, $s2$, ..., sn) be the spatial units.

Each spatial unit is associated with spatial coordinates (x_i, y_i) and micro
neighborhoods ($m1$, $m2$, ..., mi). To construct a Voronoi diagram, consider two
spatial units connected by a line segment, which is then bisected into two half
planes. The dotted lines indicate line segments connecting the spatial units. The
solid lines represent the bisecting lines, which form a Voronoi polygon surrounding
each spatial unit. The Voronoi polygon surrounding a spatial unit si is denoted by
V(si). The Voronoi polygons form a polygonal partition of the plane, called the
Voronoi diagram of the entire set of spatial units S, denoted by V(S). It is composed
of Voronoi edges and Voronoi vertices forming a polygonal cell around each spatial
unit. As new spatial units are added, more half planes are formed, and the region of
influence of the spatial unit is the intersection of the half planes. Thus, V(S) is
comprised of the entire proximity information about S in an explicit and compu-
tationally useful manner. The partitions are practically implemented using the
Fortune's sweep line algorithm [29]. Each micro neighborhood is assigned an area
Δs_j. Let Δl_j be distance between any two micro neighborhoods. In the case of a
straight tube in which the fluid speed along the trajectory is stable, the tube's
cross-section remains constant and the trajectory is populated with particles equally
spaced at distance Δl_j.

(a)   **(b)**

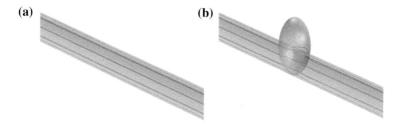

Fig. 4.12 Streamlines in tubular geometry. **a** 8 particles placed around a *circle* of radius 0.86 mm. **b** Geometry showing aneurysm with center offset by 1.5 mm from the vessel axis

During MRA simulation, the particles move along their associated trajectories towards the tube outlets, and finally disappear as they pass through the outlet. Simultaneously, new particles are introduced at the inlet of the tube. This ensures that the vascular object is filled with particles at all times of the simulation. The new particles are introduced at a specific time rate Δt_j, equal to the ratio of the distance Δl_j and speed of the fluid along the corresponding trajectory. Thus, every particle introduced at the inlet is distant from the preceding one by Δl_j, and represents the same amount of the volume equal to product of Δs_j and Δl_j. The Δl_j and Δt_j are adjusted individually for each trajectory. In a narrow path, the particles move faster and the distance Δl_j becomes larger. As a result, area of vessel cross section shrinks due to decreasing Δs_j. This is in inverse proportion to the increase of Δl_j, with the result that the volume corresponding to the particle remains constant. The particles on each trajectory belong to same streamlines, and the associated volume is used to weight the MR signal originating from the respective particle. Figure 4.12 shows the streamline plot of the trace of 8 particles uniformly placed around a circle of radius 0.86 mm at the inlet and the same with a sphere (aneurism) of radius 2 mm placed half-way along the artery, with its centre offset by 1.5 mm from the main artery axis.

The time-step of MRA simulation δt, generally differs from the step-size Δt used in flow simulation and also from the steps Δt_j used to fill trajectories with particles. Consider a sampling time-point t_{m-1} during acquisition. Let the particle be at nth position in the jth trajectory. At the next sampling instant $t_m = t_{m-1} + \delta t$, the particle has moved to the $(n + m)$th position along trajectory j. Thus the flow is charac-terized by the relative change in positions computed using respective coordinates along the trajectory. This enables performing image synthesis at arbitrary temporal resolution. Apart from volume and parameters related to their motion, the particles are assigned T_1 and T_2 relaxation times and proton density value ρ. Each particle on a trajectory that gains a unique identification label enables tracing it during image formation.

A single tubular structure is typically set to have around 256 trajectories. To minimize computations, the particle volumes are chosen to distribute the particles per mm length of the vessel, resulting in about 20–30 particles per mm^3. The number of trajectories can be reduced in low resolution images. Modeling of stationary tissue is done with set of particles whose positions are fixed for entire scope of simulation. Stationary tissues are defined by setting the shape and size of the organ. The number of stationary particles is adjusted to match the density of their distribution to the ratio obtained for the moving ones. Tissue particles are uniformly distributed within the imaged volume except for the regions occupied by vessel branches.

As discussed in Sect. 2.1.2, the magnetization update is performed using (2.19). The only difference is that the time-step used in (2.22) is now δt. For the proper evaluation of the Bloch equation, position $r_p = [x_p, y_p, z_p]^T$ of a particle is determined relative to the gradient isocentre. For a stationary component, each particle has constant coordinates. In contrast, a moving particle changes position in time. Thus, a magnetization vector is updated at time intervals which ensure reasonable trade-off between precision of the flow simulation, MR modeling and computational burden. Too high values of δt introduce discontinuities in evolution of the effective off-resonance frequency experienced by a particle. On the other hand, very short δt results in significant increase in simulation time. Typically δt is adjusted to match sampling frequency of the signal acquisition phase, i.e. it is set to sampling window duration divided by the number of k-space points in the frequency-encoding direction.

During excitation using an RF pulse of duration τ_p, the excitation interval has to be divided into time intervals δt as in other simulation phases. Though blood particles are in motion during the *RF* pulse, they do not acquire sufficient energy to get excited. At each δt step, the particle magnetization vectors are flipped only partially by angle α/n_{ts}, where $n_{ts} = \tau_p/\delta t$. In the case of tilted MRA acquisition [30], (2.25) resolves to

$$\alpha_{eff} = \tau_p \sqrt{(\omega_{RF} - \omega_0)^2 + \left(\frac{\alpha + \alpha_T(z)}{\tau_p}\right)^2} \qquad (4.21)$$

where $\alpha_T(z)$ is the extra flip angle dependent on a particle's z-position, which adds to the base angle set for the entry slice. The signal acquisition phase fills one line of k-space per *TR*. A single data point is calculated through numerical integration of transverse components of particle magnetization vectors over the whole imaging volume according to

$$s(t) = \sum_{p=1}^{n_p} \vec{M}_p(t) \cdot \hat{x} + j \sum_{p=1}^{n_p} \vec{M}_p(t) \cdot \hat{y} \qquad (4.22)$$

where n_p denotes number of particles. Sampling frequency is determined by user-defined length of the sampling time frame, and the desired number of voxels in the frequency-encoding direction. When one k-space point is acquired, the system

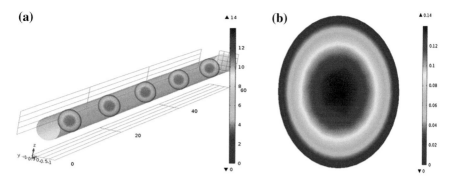

Fig. 4.13 **a** Velocity slices in a tubular geometry. **b** Velocity map for a slice located at $x = 30$ mm from origin

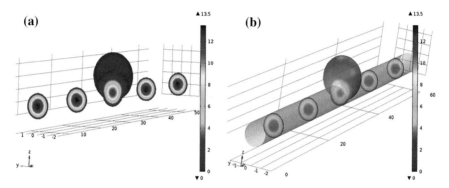

Fig. 4.14 Velocity maps for vascular geometry with aneurysm. **a** Slice: Velocity magnitude (cm/s). **b** Slice: Velocity magnitude (cm/s) Surface 1(1)

is allowed to alter its state either due to blood flow, or relaxation phenomena. After updating particle's position and their magnetization vectors, next data point is calculated. Since blood particles are subjected to flow, some of them may leave a vessel between two subsequent signal measurement times. New unexcited particles having zero transverse magnetization replace those that have disappeared. They do not produce any signal until next echo cycle. This shows the resulting image as dark voxels in the slices where blood particles of high velocity enter the FOV. Figure 4.13 shows the velocity slices computed in a tubular geometry with straight line trajectories.

For vascular geometry with aneurysm, the velocity maps are shown in Fig. 4.14. Figure 4.15 shows velocity maps at $x = 15$ mm and inside aneurysm at x = 30 mm. Once the flow field is computed, the Voronoi partitions and inlet labels are input to Bloch equation for computation of instantaneous changes in net magnetization inside the vessel.

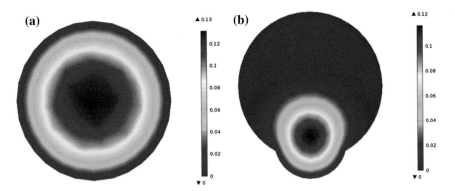

Fig. 4.15 Velocity maps for slices located **a** at $x = 15$ mm, **b** $x = 30$ mm (aneurysm) Surface: Velocity magnitude (m/s)

4.5 Bloch Flow Equations

Simulation utilizes the particle trajectories of the virtual spin packets from the time averaged numerical turbulent velocity data. The Bloch equations are solved to obtain the transverse magnetization [11, 18, 31, 32]. The particle velocity u at a given time point is computed along each trajectory, and resolved into the x, y and z components. The velocity vectors are then input into the Bloch equation where each local magnetizations are resolved along the directions of the local flow vector for updating the MR signal at each time-step.

According to Bloch equation, a particle spins with angular speed ω in a rotating coordinate system. When B_1 field is applied on a microscopic volume of mass m of red cells at equilibrium, the total force on m must be zero [33]. The forces are contact force, coriolis force and the centrifugal force [34]. The coriolis and centrifugal forces seem quite real in a rotating frame. Figure 4.16 represents the simplified model of blood vessel.

In this model, an external static field B_0 is applied along z-axis and field detector along y-axis. Once the sample reaches its equilibrium, magnetization vector M will be directed along the z-axis. By applying additional pulsed rotating field B_1 in horizontal plane, orientation of M can be shifted into this plane. Rotating coordinate system is chosen to rotate at same frequency as B_1. Bloch equation in the rotating coordinate system is expressed as

$$
\begin{aligned}
\frac{dM_x}{dt} &= V \cdot gradM_x + \frac{\partial M_x}{\partial t} = -\frac{M_x}{T_2} \\
\frac{dM_y}{dt} &= V \cdot gradM_y + \frac{\partial M_y}{\partial t} = \gamma M_z B_1(x) - \frac{M_y}{T_2} \\
\frac{dM_z}{dt} &= V \cdot gradM_z + \frac{\partial M_z}{\partial t} = -\gamma M_y B_1(x) - \frac{M_0 - M_z}{T_1}
\end{aligned}
\tag{4.23}
$$

Fig. 4.16 Cross section of an artery with longitudinal axis parallel to B_0 direction

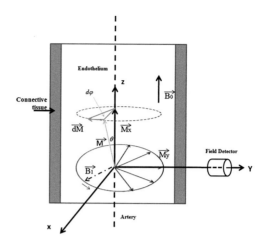

where V is the flow velocity. The Bloch equation in (4.23) can be re-arranged as

$$V^2 \frac{d^2 M_y}{dx^2} + V\left(\frac{1}{T_1} + \frac{1}{T_2}\right)\frac{dM_y}{dx} + \left(\gamma^2 B_1^2(x) + \frac{1}{T_1 T_2}\right)M_y = \frac{M_0 \gamma B_1(x)}{T_1} \qquad (4.24)$$

The transverse magnetization component M_y is calculated using two initial boundary conditions [35, 36]. (i) $M_0 \neq M_x$, a condition that holds true in general and in particular when $B_1(x)$ field is strong. This is considered as linearity condition, where frequency response takes characteristic Lorentzian form [37], (ii) resonance condition holds, (iii) fluid particles possess $M_x = M_y = 0$, before entering signal detector coil. Under the above mentioned conditions and for steady flow, $\partial M_y / \partial t = 0$. When $B_1(x)$ field is applied, M_y has largest possible amplitude when $B_1(x)$ is maximum and $M_0 \approx 0$. Thus (4.24) becomes

$$\frac{d^2 M_y}{dx^2} + \frac{T_0}{V(x)}\frac{dM_y}{dx} + \frac{1}{V^2(x)}\left(\gamma^2 B_1^2(x) + \frac{1}{T_1 T_2}\right)M_y = 0 \qquad (4.25)$$

where $T_0 = 1/T_1 + 1/T_2$. $M_y(t)$ is a simplified form of solution for the transverse magnetization. The solution can be obtained in terms of Legendre Polynomials of a given order [38]. Since the duration of excitation pulse is short compared to relaxation times, the effects of relaxation can be neglected during excitation pulse. The time-step used for solving the Bloch equation is chosen as

$$\Delta t = \frac{1}{n B_{Gmax} \gamma} \qquad (4.26)$$

where n is the number of time steps per rotation. The simulation is performed using the steps outlined in Sect. 2.2.1. Each column in the figure corresponds to the sampling points of the velocity profile in Fig. 4.17. The velocity maps for laminar

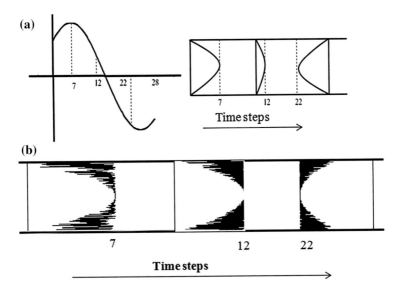

Fig. 4.17 Velocity profile at different sampling points. **a** Laminar flow. **b** Turbulent flow

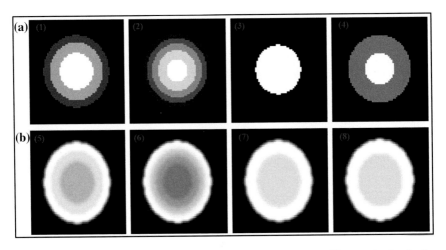

Fig. 4.18 **a** Velocity maps at different time steps for laminar flow, **b** speed images simulated using velocity maps in (**a**)

flow at different sampling points are shown in Fig. 4.18a, and simulated speed images are shown in Fig. 4.18b. The speed image simulated using turbulent flow are shown in Fig. 4.19b. Simulation is performed with 256 particles distributed in straight line trajectories.

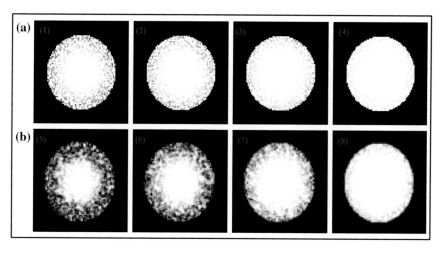

Fig. 4.19 a Velocity maps at different time steps for turbulent flow, **b** speed images simulated using velocity maps in **(a)**

References

1. Haacke EM, Brown RW, Thompson MR, Venkatesan R (1989) Magnetic resonance imaging: physical principles and sequence design. Wiley, New York
2. Cattin HB, Collewet G (2005) Numerical implementation of the Bloch equations to simulate magnetization dynamics and imaging. J MRI 173
3. Segal A (2015) Finite element methods for the incompressible Navier–Stokes equations. TU Delft
4. Taylor CA, Draney MT (2004) Experimental and computational methods in cardio vascular fluid mechanics. Annu Rev Fluid Mech 36:197–231
5. Pashaei A, Fatouraee N (2009) An analytical phantom for the evaluation of medical flow imaging algorithms. Phys Med Biol 54:1791–1821
6. Guo Z, Shi B, Wang N (2000) Lattice BGK model for Incompressible Navier-Stokes equation. J Comput Phys 165:288–306
7. He X, Luo L (1997) Lattice Boltzmann for the incompressible Navier-Stokes equation. J Stat Phys 88:3/4
8. Aspasia K (2012) Study of blood flow parameters in a phantom by magnetic resonance imaging. University of Patras, MRI School of Medicine
9. Sayah A, Mamourian AC (2012) Flow related artifacts in MR imaging and MR angiography of the central nervous system. Neurographics 2:154–162
10. Klepaczko A, Szczypiński P, Dwojakowski G, Strzelecki M, Materka A (2014) Computer simulation of magnetic resonance angiography imaging: model description and validation. PlosOne 9
11. Mikhal J (2012) Modeling and simulation of flow in cerebral aneurysms. University of Twente, Enschede
12. Jurczuk K, Kretowski M, Bellanger JJ, Eliat PA, Jalmes HS, Wendling JB (2013) Computational modeling of MR flow imaging by the lattice Boltzmann method and Bloch equation. Magn Reson Imaging 31:1163–1173

13. Jurczuk K, Kretowski M, Eliat PA, Jalmes HS, Wendling JB (2014) In silico modeling of magnetic resonance flow imaging in complex vascular networks. IEEE Trans Med Imaging 33:11
14. Tritton DJ (1988) Physical fluid dynamics. Oxford University Press, Oxford
15. Bittoun J, Taquin J, Sauzade M (1984) A computer algorithm for the simulation of any nuclear magnetic resonance (NMR) imaging method. Magn Reson Imaging 3:363–376
16. Sukop M, Thorne DT (2006) Lattice Boltzmann modeling: an introduction for geoscientists and engineers, 1st edn. Springer, Berlin
17. Koelman J (1991) A simple lattice Boltzmann scheme for Navier-Stokes fluid flow. EPL (Europhysics Letters)
18. Bao YB, Meskas J (2011) Lattice Boltzmann method for fluid simulations. www.cims.nyu.edu/~billbao/report930.pdf
19. Mazzeo MD (2009) Lattice-Boltzmann simulations of cerebral blood flow. University College London, London
20. Chen S, Martnez D, Mei R (1996) On boundary conditions in lattice Boltzmann methods. J Phys Fluids 8:2527–2536
21. He X, Luo L (1997) Lattice Boltzmann for the incompressible Navier-Stokes equation. J Stat Phys 88:3/4
22. Wolf-Gladrow DA (2000) Lattice-gas cellular automata and lattice Boltzmann models. Springer, Berlin
23. Begum R, Basit MA (2008) Lattice Boltzmann method and its applications to fluid flow problems. Euro J Sci Res 22:216–231
24. Bhatnager P, Gross EP, Krook MK (1954) A model for collision processes in gases. I. Small amplitude processes in charged and neutral one-component systems. Phys Rev 94(3):511–525
25. Guo Z, Shi B, Zheng C (2002) A coupled lattice BGK model for the Boussinesq equations. Int J Numer Meth Fluids 39:325–342
26. http://staff.polito.it/pietro.asinari/publications/PhD/AsinariDoctoralThesis_Chap5of5.pdf//a
27. Jurczuk K, Kretowski M, Eliat PA, Bellanger JJ, Jalmes HS, Wendling JB (2012) A new approach in combined modeling of MRI and blood flow: a preliminary study. IEEE, 978-1-4577-1858
28. Kwan RKS, Evans AC, Pike GB (1999) MRI simulation-based evaluation of image-processing and classification methods. IEEE Trans Med Imaging 18:1085–1097
29. Fortune S (1987) A sweep line algorithm for Voranoi diagrams Algorithmica 2:153–174
30. Wu Ex X, Hui ES, Cheung JS (2007) TOF-MRA using multi-oblique-stack acquisition (MOSA). JMRI 26:432–436
31. Sayah A (2012) Flow related artifacts in MR Imaging and MR angiography of the central nervous system. Neurographics 2:154–162
32. Peterson S (2008) Simulation of phase contrast MRI measurements from numerical flow data. Department of Biomedical Engineering, Linkoping, Sweden
33. Awojoyogbe OB (2007) A quantum mechanical model of the Bloch NMR flow equations for electrodynamics in fluids at the molecular level. Phys Scr 75:788–794
34. Awojoyogbe OB, Boubaker K (2009) A solution to Bloch NMR flow equations for the analysis of hemodynamic functions of blood flow system using m-Boubaker polynomials. Curr Appl Phys 9:278–283
35. Awojoyogbe OB, Salako KA (2005) Dynamic of the time dependent Bloch NMR equations for complex rF B1 field. International Centre for Theoretical Physics Publications
36. Awojoyogbe OB (2004) Analytical solution of the time dependent Bloch NMR equations: a translational mechanical approach. Phys A 339:437–460
37. Cowan BP (1997) Nuclear magnetic resonance and relaxation, 1st edn. Cambridge University Press, Cambridge
38. Michael DO, Bamidele AO, Adewale AO, Karem B (2013) Magnetic resonance imaging—derived flow parameters for the analysis of cardiovascular diseases and drug development. Magn Reson Insights 6:83–93

Chapter 5
Modeling of PC-MRA

Abstract This chapter highlights different post-processing techniques applied to PC-MRA speed images. Although vascular signals in PC-MRA have higher signal magnitudes as compared to background, presence of noise, eddy current and field inhomogeneities can interfere with useful information. This leads to signal losses in vascular regions, and increased signal magnitudes in background. In this situation, statistical mixture models are used to separate out the vascular signals. The chapter begins with an introduction of appropriate statistical models, and proceeds with theoretical details of carrying out vascular segmentation at both local and global levels. As part of the modeling pipeline, discussion of algorithms for vascular tree extraction and skeletonization is also presented.

Keywords Partial volume effect · Finite mixture models · EM algorithm · Skelitonization · Markov random field

5.1 An Overview of PC-MRA Modeling

The intensity projected MR images consist of artifacts generated during reconstruction process, in addition to those introduced by acquisition and processing methods prior to reconstruction. Thus, modeling of PC-MRA is essential for extracting representations of prior knowledge that can subsequently be used for artifact-free discrimination of voxels belonging to vessels and background. Modeling pipeline involves several component steps such as segmentation, geometric modeling, vascular tree extraction etc. Geometric modeling mainly involves characterizing the shape of vessel centerlines using parametric models. Several shape features based on geometric models such as centerline's local curvature and torsion are useful for comparing vessels.

Vessel lumen segmentation from angiographic image is a crucial step in PC-MRA modeling. Due to variable SNR and partial volume effects, a segmentation process normally starts with a pre-filtering of angiographic image to enhance

© The Author(s) 2016
J. Suresh Paul and S. Gouri Raveendran, *Understanding Phase Contrast MR Angiography*, SpringerBriefs in Electrical and Computer Engineering,
DOI 10.1007/978-3-319-25483-8_5

Fig. 5.1 Workflow of steps
in modeling of PC-MRA

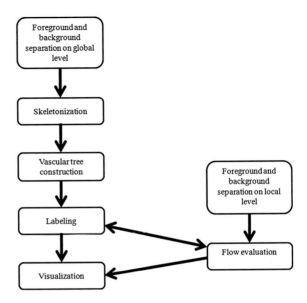

vascular, elongated structures and suppress background noise. Vascular pathologies
like stenoses and aneurysms exhibit high shape variability. Instead of modeling
them directly, segmentation methods are chosen with reference to pathologies, so
that they can be detected as deviations from the healthy vessel model.

An important step in the modeling pipeline is the extraction of vascular tree.
Vasculature, being composed of tubular shapes, can be efficiently represented by a
set of connected vessel centerline space curves. Therefore, after the segmentation, a
skeletonization procedure is performed to obtain 4D points indicating the spatial
positions of each sequential vessel skeleton point and the associated local
cross-section radius. Skeletonization methods are mainly based on topological
thinning in which the boundaries are iteratively peeled until one voxel thick medial
curve is obtained. The vascular tree is constructed from the skeleton by dividing the
centerline into branches and removing spurious centerlines that remain uncon-
nected. The branches in the vascular tree are indexed so that they are separated from
each other. In the final step in modeling, the foreground and background separation
is performed on a local scale using boundary extraction. A work flow of the steps in
PC-MRA modeling is shown in Fig. 5.1.

5.1.1 Partial Volume Effect

When the quantitative flow measurements are considered, the occurrence of error
can be systematic or random. Random error is assumed to be caused by image noise
only. The systematic error occurs due to poor vessel segmentation, partial volume

(a)   **(b)**

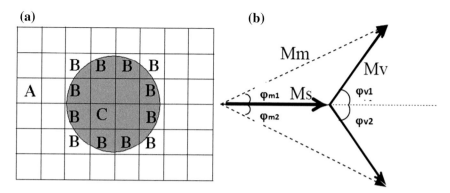

Fig. 5.2 Partial volume effect on flow measurements

effect and intra-voxel phase dispersion. Intra-voxel phase dispersion occurs at vessel voxel containing range of different velocities. Since voxel phase $\varphi = 2\pi\gamma vM_1$, the voxel phase reflects the mean phase of all spins within the voxel. The voxel magnetization is the total vectorial sum of spin magnetizations within the voxel. When spins have different range of velocities, the vectorial sum gives a result different from the normal average of spin phases. This is called intra-voxel phase dispersion error. The effect of laminar or turbulent flow on accuracy of flow measurement is negligible compared to that from partial volume effect.

Partial volume effects occur when signal generating component comes from more than one source. When flow measurements are considered, partial volume effects arise at voxels of vessel boundary which contain intra-vascular flowing spins as well as extra vascular static ones. Figure 5.2a shows a circular vessel structure and extra-vascular tissue as indicated by the gray and white colored regions. Intra-vascular voxel, 'C' are only occupied by flowing blood. Voxels 'B' consist of both flowing blood spins and extra-vascular tissue. The 'B' voxels are therefore called as partially occupied voxels. Figure 5.2b shows the partial volume effect on phase values.

Ms and Mv represent magnitudes of stationary and moving spin magnetization vectors within a voxel. Let Ns and Nv denote the number of stationary and moving spins. Subscripts 1 and 2 in Fig. 5.2b, represents positive and negative flow encodings. Let F represent the fraction of voxels occupied by flowing spins and R the ratio of modulus of moving spin magnetizations to the stationary spin magnetizations.

$$Fv = \frac{Nv}{Nv + Ns} \quad \text{and} \quad R = \frac{Mv}{Ms} \tag{5.1}$$

The tangent of measured phase φ_m is given by

$$\tan \varphi_m = \frac{FMv \sin\varphi_v}{(1-F)Ms + FMv\cos\varphi_v}$$

$$= \frac{FR\sin\varphi_v}{1 + F(R\cos\varphi_v - 1)} \tag{5.2}$$

where φ_v is phase gained by moving spins. If φ_v is small, (5.2) can be approximated as

$$\varphi_m = \frac{FM_v\varphi_v}{1 + F(R-1)} \tag{5.3}$$

Thus if stationary and moving spins have same magnetization vector magnitude or same intensity level ($R = 1$), then the measured phase will be the mean voxel phase, as opposed to phase calculated from the vectorial sum.

Once partial volume voxels are considered, the vessel area will be over estimated to compensate for reduced phase values in partial volume voxels. The impact of partial volume effect in flow calculation is negligible. Generally, moving and stationary signal intensities are different from each other and partial volume phase error increases with increasing difference in magnitude. In the case of stationary spin, $\varphi_m < \varphi_v$. As the ratio of stationary tissue within partial volume voxel increases, Ms will get stronger making φ_m smaller.

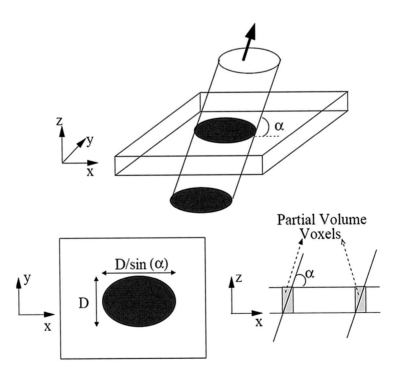

Fig. 5.3 Oblique flow direction

The partial volume effect arises due to another reason also, i.e., oblique vessel orientation with respect to flow encoding axis. This can be illustrated as shown in Fig. 5.3. Let a circular vessel of diameter D has α degree oblique orientation with respect to the imaging plane. Velocity encoding is assumed to be along the slice-select (z) axis. If the actual velocity is v, then only $v \text{Sin} \alpha$ will correspond to the measured velocity. The vessel when projected into the image plane will become an ellipse. Major axis of this ellipse is $D/\text{Sin}\alpha$ units long. Due to angulations, partial volume voxels include both stationary and moving spins. In the Fig. 5.3, in the xz cross section plane, partial volume is shown in gray. Partial volume effects resulting from vessel angulation increase with increase in slice thickness. Thicker slices result in better signal, but results in partial volume problem if angular excursion is large.

5.2 Global Segmentation of Speed Images

A speed image is considered to be composed of voxels belonging to either vessel or background. For characterizing the statistical properties of speed images, it is necessary to model the intensity or distributions for vessel and background voxels separately. Fully resolved vessels with diameters exceeding voxel dimension form major part of the high intensity tail in frequency histogram of magnitude weighted speed image. Thus, it is possible to have a comprehensive model for distribution of these voxel values that incorporates partial volume effects. Uniform distribution provides a simple estimate of distribution of fully resolved vessel signals. It is based on high speed regions, avoiding some errors associated with fitting small distribution in the presence of large competing distributions. Use of uniform distribution provides a bound on the performance of segmentation in intensity regions below a threshold determined from the use of mixture models.

There are two main reasons for introducing statistical segmentation methods to PC-MRA. First, statistical characteristics of background and vascular signals can be accurately modeled based on knowledge of image formation and physical characteristics of blood flow. Second reason arises from the use of directional flow field, available only in PC-MRA data. Since the parameters of PC-MRA model differ across voxels representative of flow and background, it is appropriate to use a Finite Mixture Model (FMM) for separating the regions at regions of low SNR. The mixture model approach to segmentation assumes that data originate from a mixture of an unknown number K of clusters in some unknown proportions $\alpha_1, \alpha_2, \ldots, \alpha_K$. The data $X = (\bar{x}_1, \bar{x}_2, \ldots, \bar{x}_n)$ are assumed to be a p-dimensional sample of size n, from a probability distribution with density

$$f(\bar{x}_j|\Theta) = \sum_{i=1}^{K} \alpha_i f_i(\bar{x}_j|\bar{\theta}_i) \qquad (5.4)$$

where the mixing probabilities satisfy $\sum_{i=1}^{K} \alpha_i = 1$, $\alpha_i \geq 0$, and $i = 1,...,K$. Each class is represented by a multivariate distribution f_i parameterized by $\bar{\theta}_i = (\mu_i, \Sigma_i)$, where μ_i is $p \times 1$ and $\sum_i$ is a $p \times p$ covariance matrix. Considering the ith distribution to be a multivariate Gaussian,

$$f_i(\bar{x}_j, \bar{\theta}_i) = \frac{1}{(2\pi)^{p/2} |\sum_i|^{1/2}} \exp\left(-\frac{1}{2} (\bar{x}_j - \mu_i)^T \sum_i^{-1} (\bar{x}_j - \mu_i) \right) \qquad (5.5)$$

The nature of component distributions in the FMM for PC-MRA is mainly dictated from definition of the speed image, and influence of noise arising from both magnitude and phase of the acquired MR data.

Since speed images are obtained by weighting the magnitude image using weights derived from the phase difference images along each direction, the noise characteristics is influenced by noise components in both magnitude and phase. The signals measured through the quadrature detector provide real and imaginary signals with an assumption that noise in each signal has a zero mean Gaussian distribution [1]. Use of Fourier reconstruction retains the Gaussian nature of noise.

Magnitude images are formed by calculating the pixel-by-pixel root Sum-of-Squares (rSoS) of the real and imaginary part of the complex image $\mathbf{Z}$. Thus $I_m = \sqrt{I_r^2 + I_i^2} = \sqrt{|Z|^2}$. The noise distribution in the magnitude I_m will not be Gaussian due to the squaring operation. Denoting the image pixel intensity in the absence of noise by I_{an}, and measured pixel intensity by I_m, the probability distribution of measured intensity is Rician distributed for a random variable I_m [2].

$$f(I_m) = \frac{I_m}{\sigma^2} \exp\left(-(I_m^2 + I_{an}^2)/2\sigma^2 \right) I_0 \left(\frac{I_{an} \cdot I_m}{\sigma^2} \right) \varepsilon(I_m) \qquad (5.6)$$

where I_0 is the modified 0th order Bessel function of the first kind [3] and σ is the standard deviation of Gaussian noise in the real and imaginary images. The unit step Heaviside function $\varepsilon(.)$ indicates that the expression for pdf is valid for non-negative I_m values only. The shape of pdf (5.6) depends on SNR defined as I_{an}/σ. In regions where $I_{an} = 0$, (5.6) reduces to a special case of Rician distribution, known as Rayleigh distribution [4, 5]. In this case, the measured intensity distribution is given by

$$f(I_m) = \frac{I_m}{\sigma^2} \exp\left(-(I_m^2)/2\sigma^2 \right) \varepsilon(I_m) \qquad (5.7)$$

For increasing value of SNR, $\chi = \frac{I_{an} I_m}{\sigma^2}$ will be large. Using the asymptotic expansion of $I_0(\chi)$, it can be shown that $I_0(\chi) \approx \frac{e^\chi}{\sqrt{2\pi\chi}}$ [6]. In this case, the Rician distribution can be approximated as

$$f(I_m) = \sqrt{\frac{I_m}{2\pi\sigma^2 I_{an}}} \exp\left(-\frac{(I_m - I_{an})^2}{2\sigma^2}\right)$$
$$\approx \frac{1}{\sigma\sqrt{2\pi}} \exp\left(-\frac{(I_m - I_{an})^2}{2\sigma^2}\right) \tag{5.8}$$

Therefore, it can be inferred that the Rayleigh distribution governs the noise in image region with no MR signal. The analytically evaluated mean and variance of the distribution in (5.7) will be

$$\bar{I}_m = \sigma\sqrt{\pi/2} \quad \text{and} \quad \sigma_{I_m}^2 = (2 - \pi/2)\sigma^2 \tag{5.9}$$

These relations give the noise power σ^2 from the magnitude image. For high SNR regions, the mean and variance are approximately I_{an} and σ^2. The phase images are reconstructed from real and imaginary images by calculating the arctangent of their ratio pixel-by-pixel. The PDF of complex Gaussian random variables $\mathbf{z}$ with mean μ and variance σ^2 is

$$f(z) = \frac{1}{2\pi\sigma^2} \exp\left(-\frac{|z-\mu|^2}{2\sigma^2}\right) \tag{5.10}$$

When $\mathbf{z}$ is expressed in polar coordinates as $z = I_m e^{j\varphi_m}$, the complex mean will be $\mu = I_{an} e^{j\varphi_{an}}$. The distribution in (5.10) can now be written in the form of joint distribution as

$$f(I_m, \varphi_m) = \frac{1}{2\pi\sigma^2} \exp\left(-\frac{|I_m e^{i\varphi_m} - I_{an} e^{i\varphi_{an}}|^2}{2\sigma^2}\right) \tag{5.11}$$

As discussed above, the magnitude is Rician distributed as given in (5.6). The conditional distribution of the phase for a given magnitude is obtained by dividing (5.11) by (5.6) and takes the form of a Tikhonov distribution [7, 8].

$$f(\varphi_m, I_m) = \frac{\exp(\lambda\cos(\varphi_m - \varphi_{an})}{2\pi I_0(\lambda)} \tag{5.12}$$

where $\lambda = I_m I_{an}/\sigma^2$. The marginal pdf of phase, obtained by integrating (5.11) over I_m [5, 9, 10] is

$$f(\varphi_m) = \frac{1}{2\pi} \exp\left(-\frac{1}{2}\left(\frac{I_{an}}{\sigma}\right)^2\right)\left(1 + \kappa\sqrt{\pi}\exp(\kappa)^2(1 + erf(\kappa))\right) \tag{5.13}$$

where $\kappa = (1/\sqrt{2})\,(I_m/\sigma)\cos(\varphi_m - \varphi_{an})$. For zero SNR, $I_{an} = 0$ and both conditional and marginal PDF of phase reduces to that of a uniform PDF as in

$$f(\varphi_m|I_m) = f(\varphi_m) = {}^1\!/_{2\pi} \qquad (5.14)$$

Similarly, for high SNRs,

$$f(\varphi_m) = \frac{1}{\sigma\sqrt{2\pi}}\exp\left(\frac{-I_{an}^2(\varphi_m - \varphi_{an})^2}{2\sigma^2}\right) \qquad (5.15)$$

The distribution of phase noise is shown in Fig. 5.4. Gaussian approximation is good even for small SNR. In some cases, the phase images are weighted by magnitude data to reduce the phase variations in region with no signal. The SNR is calculated as M_1/σ, where M_1 is the magnitude of the signal and σ is the standard deviation of noise.

The effect of noise on the phase angles is modulated by the signal magnitude, with the assumption that for lower signal intensities, the distribution of phase difference will be more uniform. $\Delta\varphi_x, \Delta\varphi_y, \Delta\varphi_z$, is assumed to follow zero-mean Gaussian distribution. The modulus of three independent zero-mean Gaussians with equal variance describes the Maxwell distribution [11]. The pdf of background signal follows a Maxwell distribution

$$f_M(i) = \frac{2}{\sqrt{2\pi}}\frac{i^2}{\sigma_M^3}\exp\left(\frac{-i^2}{2\sigma_M^2}\right) \qquad (5.16)$$

where $i \geq 0$. Thus overall pdf of background and vessel can be a Maxwell-Uniform (MU) mixture model

$$f_{MU}(i) = w_M f_M(i) + w_U f_U(i) \qquad (5.17)$$

Fig. 5.4 Distribution of phase noise

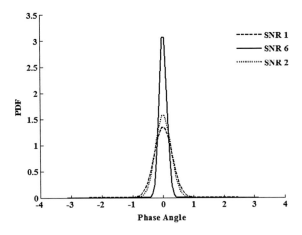

Fig. 5.5 Histogram showing tail end not perfectly fitted by Gaussian distribution

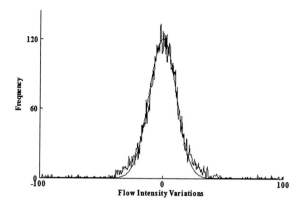

The deviation of magnitude weighted phase values $\vec{M}\Delta\varphi_j, j = x, y, z$ from a Gaussian distribution results in the failure of Maxwell distribution in describing the background signals. As shown in Fig. 5.5, tails of the magnitude-weighted phase image histogram cannot be properly fitted by a Gaussian when SNR is low. Since there is Rician nature for noise in the signal magnitude [11], the distribution of their product, the magnitude weighted phase value is not entirely Gaussian.

To reduce the cause of failure, two small non-zero mean Gaussian residual distributions are introduced. Thus, for each encoding direction, PDF of $\vec{M}\Delta\varphi$ consists of a zero mean Gaussian and two non-zero mean Gaussian distributions. After modulus operation, Maxwell distribution is formed by the modulus of three zero-mean Gaussian distributions whereas modulus of residual non-zero mean Gaussian distributions result in Gaussian distributions [9, 12]. Therefore, the PDF of background signal consist of a linear mixture of Maxwell distribution $f_M(i)$ with variance σ_M^2 and Gaussian distribution $f_G(i)$ with mean μ_G and variance σ_G^2. With laminar flow, the vascular signal has a uniform distribution. Thus the overall PDF of a PC-MRA speed image can be modeled as a summation of Maxwell–Gaussian and uniform finite mixture distribution given by

$$f_{MGU}(i) = \underbrace{w_M f_M(i) + w_G f_G(i)}_{\text{Background signal}} + \underbrace{w_U f_U(i)}_{\text{Vascular signal}} \qquad (5.18)$$

The Gaussian component of the MGU mixture model has a relatively large variance and small weight. The choice of MU or MGU model to fit the histogram of a given speed image is decided based on the relative entropy which is widely used for measuring the distance between two distributions. Given two distributions f and g, the relative entropy of f with respect to g, also called the Kullback–Leibler divergence (KLD) is defined by

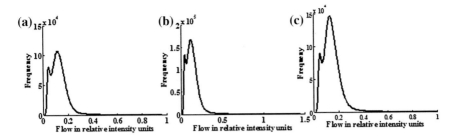

Fig. 5.6 PC-MRA Volume histogram generated from 3 different data sets

$$J(f|g) = \sum_i f(i) \log_2 \left(\frac{f(i)}{g(i)} \right) \tag{5.19}$$

When MU model fits well to the intensity histogram, added Gaussian component in the MGU model has relatively large variance, and shifts to the high intensity range of histogram. Consequently, the image can be under segmented due to the relatively high threshold. Though both histograms look identical, their relative entropies satisfy the condition

$$J_2(\bar{w}_M\bar{f}_M \| w_M f_M) < J_1(\bar{w}_M\bar{f}_M + \bar{w}_G\bar{f}_G \| w_M f_M) \tag{5.20}$$

When MU model does not fit well to the intensity histogram, the added Gaussian component in fitted MGU model has relatively small variance, and shifts to low intensity region in histogram. In this case, the relative entropies satisfy the condition

$$J_1(\bar{w}_M\bar{f}_M + \bar{w}_G\bar{f}_G) \| w_M f_M) < J_2(\bar{w}_M\bar{f}_M \| w_M f_M) \tag{5.21}$$

These conditions are followed in practice to check which of the two mixture models should be chosen to model the observed histogram.

Figure 5.6 represents the volume histogram generated from speed images of the three volunteer data sets discussed in Chap. 3. The volume histograms are constructed from the speed images from each slice computed using CD method. It is observed that the volume histogram generated has a peak near the low intensity region, which shows that PC-MRA suppresses stationary tissues [13].

5.3   Initial Estimation of Mixture Parameters

Posterior probability of group membership in the mixture model for each voxel is estimated using the knowledge of each density function. Parameter of each density function is estimated using the EM algorithm described later. To delineate the acquired data into tissue and background, this algorithm assigns each voxel to the group for which its membership has the highest posterior probability.

An initial estimation of the parameters is obtained from the shape of the observed histogram h. For conceptualizing the mixture model, h is partitioned into component histograms h_M^{init} and h_G^{init} representing the Maxwell and Gaussian histograms respectively. The initial histogram of the Maxwell distribution is treated as $h_M^{init}(i) = Cf_M(i|\sigma_M^{init})$, where the scaling factor C is set to make amplitude of $h_M^{init}(i)$ same as that of $h(i)$. Here, σ_M^{init} is determined from the peak intensity (I_{peak}) corresponding to the maximum frequency of the observed histogram. Let A_M denote area of $h(i)$ covered by $h_M^{init}(i)$, and A_{total} the total area covered by $h(i)$. The mixture ratio w_M^{init} is then determined as ratio of A_M/A_{total}. After initializing the Maxwell part, the residual histogram is computed for the bin centers greater than I_{peak} using

$$h_{res}(i) = abs(h(i) - h_M^{init}(i)) \tag{5.22}$$

Then μ_G^{init} and σ_G^{init} of the Gaussian part are determined using the first and second order moments of bin centers in the residual histogram. The initial Gaussian part is now modeled as $h_G^{init}(i) = C'f_G(i|\mu_G^{init}, \sigma_G^{init})$, with the scaling factor C' set to make amplitude of $h_G^{init}(\mu_G^{init})$ same as that of $h_{res}(\mu_G^{init})$. If A_G denote the area of $h_{res}(i)$ covered by $h_G^{init}(i)$, the initial mixture weight w_G^{init} of the Gaussian part is defined as the ratio of A_G/A_{total}. The initial component of uniform distribution in the mixture model is set to be $f_U^{init}(i) = \frac{1}{I_{max}}$, where I_{max} is the maximum intensity in the observed frequency histogram. The initial mixture weight for the uniform distribution part is then obtained as $w_U^{init} = 1 - \left(w_M^{init} - w_G^{init}\right)$.

Once parameters of the initial mixture model are determined, the initial labels for vessel and background pixels are assigned based on the MAP criterion. Using the likelihood functions and mixture weights computed from the observed histogram, a label '1' is assigned to a voxel if $w_U^{init}f_U^{init}(i) > w_M^{init}f_M^{init}(i) + w_G^{init}f_G^{init}(i)$ and '0' otherwise. According to MAP criterion, the image threshold I_{TH} is defined as the intersection of Maxwell–Gaussian and Uniform distributions. Consequently, I_{TH} can be set to a value that satisfies

$$w_M^{init}f_M^{init}(I_{TH}) + w_G^{init}f_G^{init}(I_{TH}) = \frac{w_U^{init}}{I_{max}} \tag{5.23}$$

5.3.1   Iterated EM Algorithm

Following an initial estimation of parameters as in Sect. 5.3, the EM algorithm [14] is used iteratively for re-estimating the parameters until convergence is obtained. EM Algorithm involves repeated application of Baye's rule with a constraint of maximizing the likelihood functions. Given $f(i|\Theta) = \sum_{j=1}^{K} \alpha_j f_j(i|\bar{\theta}_j)$, the EM algorithm consists of the following steps computed iteratively.

Step 1:   Start with the observed histogram $h(i)$

Step 2:   Compute the initial estimates $h_j^{(0)}(i) = C_j f_j(i|\bar{\theta}_j^{(0)})$, with $\bar{\theta}_j^{(0)}$ and $x_j^{(0)}$
          determined as discussed in Sect. 5.3

Step 3:   Compute $P_j(\bar{\theta}_j^{(0)}|i) = \dfrac{\alpha_j^{(0)} f_j(i|\bar{\theta}_j^{(0)})}{f(i|\Theta_j^{(0)})}$, where $f(i|\Theta_j^{(0)}) = \displaystyle\sum_{j=1}^{K} \alpha_j^{(0)} f_j(i|\bar{\theta}_j^{(0)})$

Step 4:   Set k = 1, and compute $\alpha_j^{(k)} = \dfrac{1}{N} \displaystyle\sum_{i=1}^{N} h(i) P_j(\bar{\theta}_j^{(k-1)}|i)$

Step 5:   Compute $\mu_j^{(k)} = \dfrac{\sum_{i=1}^{N} i h(i) P_j(\bar{\theta}_j^{(k-1)}|i)}{\sum_{i=1}^{N} h(i) P_j(\bar{\theta}_j^{(k-1)}|i)}$

Step 6:   Compute $\sigma_j^{2\,(k)} = \dfrac{\sum_{i=1}^{N} \left(i - \mu_j^{(k)}\right)^2 h(i) P(\bar{\theta}_j^{(k-1)}|i)}{\sum_{i=1}^{N} h(i) P_j(\bar{\theta}_j^{(k-1)}|i)}$

Step 7:   Set $\bar{\theta}_j^{(k)} = [\mu_j^{(k)}, \sigma_j^{2\,(k)}]$

Step 8:   Compute $f_j(i|\bar{\theta}_j^{(k)})$ and repeat steps 3 to 7 until the likelihood functions
          $L_j(\bar{\theta}_j^{(k)}|i) - L_j(\bar{\theta}_j^{(k-1)}|i)$ is less than a preset tolerance.

The iterative procedure maximizes log likelihood of the mixture distribution. The Log-likelihood function is defined as

$$L^k = \sum_{i=0}^{I_{max}} h(i) \log f(i|\Theta_j^{(k)}) \qquad (5.24)$$

Figure 5.7 shows the Log-likelihood as a function of iterations for the three vol-unteer PC-MRA data sets. The change in log-likelihood function during each successive iterations is given by

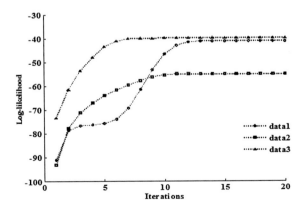

Fig. 5.7 Log likelihood as a function of iterations for data sets *1–3*

$$L^{k+1} - L^k = \sum_{i=0}^{I_{max}} h(i) \log \left(\frac{f^{k+1}(i)}{f^k(i)} \right) \qquad (5.25)$$

5.3.2   Segmentation Using Local Phase Coherence

The local phase coherence (LPC) measure is derived to estimate the degree of coherence among the neighboring voxels. It provides useful information about the flow coherence within a local region. For segmentation, the LPC measures can be used to distinguish locally coherent flow fields from (non-coherent) random flow fields. The formal definition of the LPC measure is based on the dot product of the neighboring flow vectors within a window. Using the mask centered at s shown in Fig. 5.8, LPC is computed using

$$LPC(s) = P(\bar{v}_1, \bar{v}_2) + P(\bar{v}_2, \bar{v}_3) + \ldots + P(\bar{v}_8, \bar{v}_1) \qquad (5.26)$$

where $P(\bar{v}_i, \bar{v}_j)$ denotes dot product of $\bar{v}_i$ and $\bar{v}_j$. $\bar{v}_i$ shown inside the mask contains phase differences $\Delta\varphi_x$, $\Delta\varphi_y$ and $\Delta\varphi_z$ measured at the jth position as its elements. Segmentation is performed by treating voxels with LPC value exceeding a threshold value ($P_{TH} > 0.9$) as vessels and others as background.

5.3.3   Segmentation Using MRF Formulation

Using the Markov random field (MRF) model, MAP estimation is performed using the local relationships between image voxels. In MRF based segmentation, the first step is to classify voxels in the speed image as vessel or background, with an assumption that prior probability $P(x_s = v$ or $b)$ is a constant using Maximum-a posteriori (MAP) method. The LPC measure and the present state of each voxel are

Fig. 5.8 Neighboring velocities of voxel s

$\bar{v}_1$	$\bar{v}_2$	$\bar{v}_3$
$\bar{v}_8$	$\bar{v}_s$	$\bar{v}_4$
$\bar{v}_7$	$\bar{v}_6$	$\bar{v}_5$

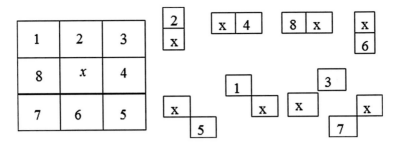

Fig. 5.9 Illustration of 8–two-site cliques

treated as prior knowledge to define prior probabilities of the vessel and background. The prior probability is given by the Gibbs function [15], with the assumption that each voxel state depends on its neighbors in an in-plane 3×3 neighborhood. A second order neighboring system is defined with 8 two site cliques. Figure 5.9 illustrates the cliques for a 3×3 mask labeled x. The prior probability $P(x_s)$ and the total energy function $U(x_s)$ for voxel s are defined as

$$P(x_s) = Z^{-1} \exp\{U(x_s)\} \text{ and } U(x_s) = \sum_{c=1}^{8} E_c(x_s) \qquad (5.27)$$

where $E_c(x_s)$ denotes the clique energy function, and $U(x_s)$ is used to define the Gibbs function $\exp(U(x_s)) \cdot Z = \sum_{s} \exp\{U(x_s)\}$ is a normalization constant.

The interactions among voxels are measured by a clique energy function $E_c(x_s)$ which promotes LPC homogeneity and assigns lower probabilities to the voxels far away from the vasculature. The clique energy functions for the vessel and background voxels are chosen as

$$E_c(x_s = v) = \begin{cases} 1 \text{ if } x_c = v \text{ and } x_s \text{ and } x_c \text{ are coherent,} \\ 0 \text{ otherwise} \end{cases}$$

$$\qquad (5.28)$$

$$E_c(x_s = b) = \begin{cases} 0 \text{ } f \text{ } x_c = v \text{ and } x_s \text{ and } x_c \text{ are coherent,} \\ 1 \text{ otherwise} \end{cases}$$

From (5.28), the Markov prior probability of vessel $P(x_s = v)$ is directly proportional to the multiplier of the number of adjacent and coherent vessel voxels. The iterated conditional modes (ICM) [16] algorithm is employed to maximize the probability of the true segmented binary image given the observed image, by changing the tissue type at each voxel according to the LPC measure and current tissue type of the neighboring voxels. The speed images generated for each data set are processed using MRF model prior to MIP reconstruction. Figure 5.10 represents the MIP images before and after application of segmentation process.

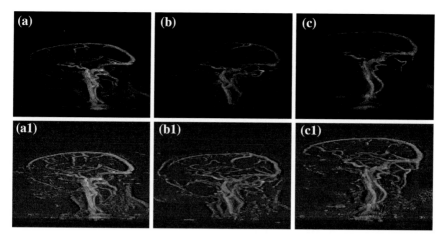

Fig. 5.10 MIP images **a–c** before segmentation **a1–c1** after segmentation

5.4 Vascular Tree Construction

A vascular tree construction provides explicit structural information about the vascular network which is not apparent in the raw images. The construction is performed by finding the skeleton of the voxels labeled as "vessels" by a process called skeletonization. The skeleton allows for separation and indexing of branches by detecting bifurcations. The vascular tree construction represents each branch in the vascular network as a one voxel thick curve in 3D space. The construction also allows for the calculation of parameters such as blood flow and geometry separately for each branch. Further, these parameters can be evaluated within a branch along the curve. Vascular tree construction is performed by an iterative algorithm named **vascular tree construction algorithm** [17] described below.

```
while branches are being removed do
        separate skeleton into branches
        calculate length of branches
        for all branches do
            if branch too short then
                    remove branch
                    update branch removal counter
            end
        end
end
```

5.4.1   Skeletonization

A skeleton is a representation of the shape in a binary image, and represents the
basic structure of an object in a topology conserving way. An approximation to a
topology conserving skeleton can be produced using thinning techniques. Thinning
is a technique that works iteratively to reduce an object in size, peeling of
layer-by-layer until only an approximation to the skeleton remains. The thinning
technique used for PC-MRA [18] is suitable for long objects such as blood vessels.
After the global segmentation step, each element in a discrete binary 3D image X
will have values of 1 or 0 representing either a vessel or background. The set $V \subseteq X$
represents all vessel points and X\V the set of all background points in the speed
image.

Figure 5.11 shows a set of points $N_j(p)$, $j = 6, 18, 26$ being an adjacency
parameter with reference to a point p. Point p is 6-adjacent to the points marked U,
D, N, S, E, and W; 18-adjacent to the points marked U, D, N, S, E, W and ●; and
26-adjacent to the points marked U, D, N, S, E, W, O and ●. A set of vessel or
background points in $N_j(p)$ is j-connected if a continuous path can be drawn through
the entire set of points using j-adjacencies. In other words, in any subset of points
smaller than that entire set, at least one of those points needs to be j-adjacent to a
point not included in that subset. Deleting points in the image X is the process of
morphing vessel points into background points. A vessel point is a simple point, if
its deletion does not change the image topology. Simple points are deleted to
produce the skeleton, and are found by investigating the connectivity between
vessel and background points in a $3 \times 3 \times 3$ neighborhood. A vessel point 'p' is a
simple point only if the following conditions hold:

1. The set $N_{26}(p) \cap V \setminus \{p\}$ is not empty
2. The set is 26-connected in itself

N6(p)=points U,D,N,S,E,W
N18(p)=N6(p)+ points marked ●
N26(p)=N18(p)+ points marked ○

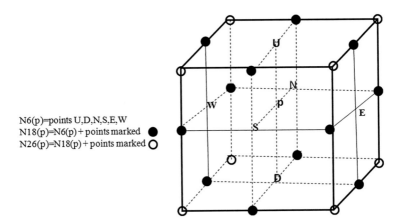

Fig. 5.11 Cube of adjacencies

3. The set $X\backslash V \cap N_6(p)$ is not empty
4. The set $X\backslash V \cap N_6(p)$ is 6 connected in the set $X\backslash V \cap N_{18}(p)$.

The first condition makes sure that the point p is not isolated. If it is isolated, it should not be deleted. The third condition tests if the point p is a border point. If not, it should not be deleted. Conditions 2 and 4 test the connectivity of vessel and background points. The third condition is evaluated in the U, D, N, S, E, and W directions sequentially according to the thinning algorithm. This means that vessel points that are border points in the U direction will be deleted first. Vessel borders being points in the D direction will be deleted second, and so on. The algorithm is sequential, first marking points for deletion in one direction at a time, followed by a re-test of the four conditions to delete simple points. The algorithm works iteratively, deleting one layer of border points at a time until no simple points remain to be deleted. The final output is the thinned skeleton; Y. Considering that simple points are deleted sequentially in the U, D, N, S, E, and W directions, skeletons can be produced with some variation depending on the order in which the conditions are tested. The thinning algorithm is summarized below.

```
Set Y = X

while vessel points are being deleted do
        delete simple(Y; U)
        delete simple(Y; D)
        delete simple(Y; N)
        delete simple(Y; S)
        delete simple(Y; E)
        delete simple(Y; W)
end

function delete simple(Y, dir)
for all points p in Y do
        test the four conditions
        if point p is simple in the dir direction then
                        mark point p as deletable
        end
end
for all deletable points do
        re-test the four conditions
        if simple then
                delete point
        end
end
```

The skeleton produced by the method described above is a one-voxel thick representation of the vascular tree. This representation enables separation of branches from each other, by identifying end points and junction points. A skeleton point p will have a branching index m depending on how many other skeleton points are there in the neighborhood $N_{26}(p)$. Based on the value of m, p is classified as singular point, end point, middle point and junction point for $m = 0, 1, 2$ and $>=3$ respectively. Using the above classification, a branch in the vascular tree corresponds to a 26-connected set of middle points. Similarly, a junction in the vascular tree corresponds to a 26-connected set of junction points. When thinning algorithm is implemented, the four conditions stated above are tested inside the while loop. One layer of border points is deleted for each iteration of the while loop. The *delete simple* function is sequential in that it first marks simple points for deletion, and then deletes the marked points one by one if the conditions still hold.

An undesired effect of the thinning algorithm is that it creates spurs [17]. A spur is a spurious part of the skeleton that does not represent a real branch. It is a short object originating from the true skeleton, stretching out towards the wall of a blood vessel. The process of removing spur is called pruning. The pruning criteria also remove isolated sticks, a term used to refer skeleton segments not connected to a larger skeleton object. A branch is a spur if it is shorter than 8 voxels and the branch ends in at least one end point. The first criterion is designed to remove any possible spurs, but could also remove some true peripheral branches. The second criterion ensures that any short intermediate branches are not removed. In the final representation, each branch, or each segment, is indexed with numbers ranging from $i = 1......Nsegments$, where *Nsegments* is the total number of segments in the vascular tree construction. The points in a segment are further indexed so that for a branch bi consisting of points $bi = \{p_{i,1}...,p_{i,Ni}\}$, the points $p_{i,1}$ and $p_{i,Ni}$ are the start and end points of the segments.

References

1. Riberio MI (2004) Gaussian probability density functions: properties and error characterization. Institute for Systems and Robotics, Technical Report Portugal
2. Rice SO (1944) Mathematical analysis of random noise. Bell Syst Tech 23:282–332
3. https://math.uc.edu/~halpern/diffeqns/Handouts/Besselforode.pdf
4. Gudbjartsson H, Patz S (1995) The Rician distribution of noisy MRI data. Magn Reson Med 34:910–914
5. Sijbers J, den Dekker AJ, Audekerke JV, Verhoye M, Van Dyck D (1998) Estimation of the noise in magnitude MR images. Magn Reson Imaging 16:87–90
6. Abramowitz M, Stegun IA (1970) Handbook of mathematical functions. Dover Publications, New York
7. Fu H, Kam P (2008) Exact phase noise model and its application to linear minimum variance estimation of frequency and phase of noisy sinusoid. In: Proceedings of the IEEE 19th international Symposium on personal indoor mobile radio communication (PIIMRC) 15
8. Donoughue NO', Moura J (2012) On the product of independent complex Gaussians. IEEE Trans Signal Process 60(3):1050–63
9. Lathi BP (1983) Modern digital and analog communication systems. Hault-Saunders International Edition, Japan
10. Bonny JM, Renou JP, Zanca M (1996) Optical measurement of magnitude and phase from MR data. J Magn Reson 113:136–144
11. Andersen AH, Kirsch JE (1996) Analysis of noise in phase contrast MR imaging. Med Phy 23:857–869
12. Chung ACS (2001) Vessel and aneurysm reconstruction using speed and flow coherence information in phase contrast magnetic resonance angiograms. Department of Engineering Science, University of Oxford, Trinity

13. Sabry M, Sites CB, Farag AA, Hushek S, Moriarty T (2002) Statistical cerebrovascular segmentation for phase contrast MRA data. University of Louisville, Louisville acbme, pp 32–37
14. Bilmes JA (1998) A gentle tutorial of the EM algorithm and its application to parameter estimation for gaussian mixture and hidden Markov models. Department of Electrical Engineering and Computer Science, Berkeley
15. Poisson E (2000) Statistical physics II. Department of Physics, University of Guelph, Guelph
16. Besag J (1986) On the statistical analysis of dirty pictures. J R Stat Soc (B) 48(3):259–302
17. Chen Z, Molloi S (2003) Automatic 3D vascular tree construction in ct angiography. Comput Med Imaging Graph 27(6):469–479
18. Palagyi K, Sorantin E, Balogh E, Kuba A, Halmai C, Erd-ohelyi B, Hausegger K (2001) A sequential 3D thinning algorithm and its medical applications. In Proceedings of the international conference on information processing in medical imaging (IPMI). Lecture notes in computer science, vol 2082. Springer, Berlin, pp 409–415

Appendix

A.1   List of Codes

Chapter 2

1. Main function for Gradient echo simulation

   (a) mrirecon_gradechoV2_dMBydT.m

2. Calling functions in GRE simulation

   (a) Kspace_gradechoV2_dMBydT.m
   (b) sliceout_gradechoV2_dMBydT.m
   (c) gradecho_wgradV2_dMBydT.m

3. Main function for Spin echo simulation

   (a) mrirecon_spinechoV2_dMBydT.m

4. Calling functions in spin echo simulation

   (a) Kspace_spinechoV2_dMBydT.m
   (b) sliceout_spinechoV2_dMBydT.m
   (c) spinecho_wgradV2_dMBydT.m

5. Simulation of SWI with air gap

   (a) simulateswiwithairgap.m

6. Simulation of SWI without air gap

   (a) simulateswiwithoutag.m

© The Author(s) 2016
J. Suresh Paul and S. Gouri Raveendran, *Understanding Phase Contrast
MR Angiography*, SpringerBriefs in Electrical and Computer Engineering,
DOI 10.1007/978-3-319-25483-8

Chapter 3

1. Computation of complex difference speed images

   (a) CD Speed.m

2. Computation of phase difference speed images

   (a) PDspeed.m

3. Main function of higher dimensional Homodyne filter for PC-MRA

   (a) Homodyne3D.m

4. Calling functions for higher dimensional homodyne filter for PC-MRA

   (a) volumerecon.m

Chapter 5

1. Function for Log-Likelihood computation using Expectation—Maximization algorithm

   (a) LL_EM.m